PIC Projects

PIC Projects

A Practical Approach

Hassan Parchizadeh
and
Branislav Vuksanovic
University of Portsmouth, UK

WILEY

John Wiley & Sons, Ltd

Telephone (+44) 1243 779777

Email (for orders and customer service enquiries): cs-books@wiley.com

Visit our Home Page on www.wiley.com

Cover image by Victor L. Dunn, University of Portsmouth, reproduced with permission

Other Wiley Editorial Offices

John Wiley & Sons Inc., 111 River Street, Hoboken, NJ 07030, USA

Jossey-Bass, 989 Market Street, San Francisco, CA 94103-1741, USA

Wiley-VCH Verlag GmbH, Boschstr. 12, D-69469 Weinheim, Germany

John Wiley & Sons Australia Ltd, 42 McDougall Street, Milton, Queensland 4064, Australia

John Wiley & Sons (Asia) Pte Ltd, 2 Clementi Loop #02-01, Jin Xing Distripark, Singapore 129809

John Wiley & Sons Canada Ltd, 6045 Freemont Blvd. Mississauga, Ontario, L5R 4J3 Canada

Wiley also publishes its books in a variety of electronic formats. Some content that appears in print may not be available in electronic books.

Library of Congress Cataloging-in-Publication Data

Parchizadeh, G. Hassan.

   PIC projects: a practical approach/Hassan Parchizadeh and Branislav Vuksanovic.

     p. cm.

   Includes bibliographical references and index.

   ISBN 978-0-470-69461-9 (pbk.: alk. paper) 1. Programmable controllers.
2. Microcontrollers–Programming. I. Vuksanovic, Branislav. II. Title.

   TJ223.P76P36 2009

   629.8'95—dc22   2009001885

British Library Cataloguing in Publication Data

A catalogue record for this book is available from the British Library

ISBN 978-0-470-69461-9 (pbk)

Typeset in 11/13pt, Sabon-Roman by Thomson Digital, India

Contents

About the Authors

Hassan Parchizadeh is a principal lecturer in the Department of Electronic and Computer Engineering at the University of Portsmouth. He has carried out various industrial projects and consultancies and has run short industrial courses. He has been lecturing for more than 20 years in topics covering analogue and digital electronics, microcontrollers, power electronics, signal processing and C programming. This is his third book in the area of digital electronics and microcontrollers.

Branislav Vuksanovic is a senior lecturer in the Department of Electronic and Computer Engineering at the University of Portsmouth. Previously, he lectured at the University of Derby and worked as a research fellow at Sheffield and Birmingham. He is currently teaching and has published a number of research papers in the areas of digital signal processing, active noise control, medical image processing and data compression.

Preface

After teaching digital electronics and microprocessors programming at university for a number of years, we noticed a shift in the way this subject is taught and in the students' expectations of this subject. Students do not like to read theory or long explanations about various types of microprocessors/microcontrollers and their particular features. Instead, they want 'hands-on' experience as early as possible. 'Getting something working' is their goal. After that initial stage and (hopefully) some success they are more willing to try to explore the field further.

During the introductory lectures we usually recommend a number of books to students, suitable for their particular courses. The majority of our students rarely read those books in any depth. Long introductory chapters and explanations about various features of some particular type of microprocessor/microcontroller tend to 'turn them off'. Most of them become alive and interested in the subject again only in the laboratories. This happens because in the laboratory they are faced with practical problems – 'how can I get that #*@**! thing working?' or 'where did I go wrong?'. Faced with these problems, students tend to ignore the books and lecture notes and use only one place to search for instant answers to their problems – the Internet. The reason is clear after looking at the various Internet sites. Most of the relevant sites are written by 'hobbyists' who do not tend to delve deeply into the theory of the specific device, but do come up with straight answers to student problems: practical, working programs, or at least programs that are supposed to work properly.

Our book takes this idea of instant answers to students' problems a number of steps beyond the random web site with a PIC program attached to it. First, all of the programs provided in the book are fully tested and proven to be working properly with specific software and hardware support. Our experience is that many of the PIC programs found on the Web do not compile or assemble cleanly. Even if this initial stage is passed, many programs still fail to operate properly when downloaded on a specific microcontroller device. There are many reasons. Some of those programs are out of date, written for earlier or even non-existent versions of supporting software (such as compilers or assemblers) or hardware. Often, information about which compiler should be used to compile the program is not even mentioned on the Internet sites. Programs are frequently poorly written and contain errors. Similar problems exist with circuits to test those programs.

PIC Projects provides two versions of the programs for most projects in the book: assembler and C. Students can benefit from and understand the principles behind the program operations by studying both versions. Programs in Chapters 5 and 6 have no assembler versions; only C language versions are supplied. The task of producing the assembler versions is left to the reader, who should at that stage be able to write his own assembler programs for those two chapters. This will certainly be a good exercise and will improve readers' knowledge of assembler language significantly.

Each project described in the book is supported by detailed explanations to help readers produce circuits, assemble or compile programs and test programs easily and see them working fully on the circuit. Circuits and programs become more complex as the reader goes through the book but most of them should be within the capabilities of first- or second-year university students. Various features and members of the PIC family are introduced gradually: PIC12F629, PIC12F675, PIC16F627A and PIC18F252 are all used for various projects but none of the features of these devices is discussed in great detail unless this is required by the specific project in the book. Instead, it is left to the reader to research further about a particular feature.

This book is also intended for a college or university lecturer who needs to organize a new course in an area related to digital electronics or who wants ideas for lab exercises in his existing unit. Exercises in the first part of the book will be sufficient as a starting point for undergraduate PIC based courses while the last two chapters contain some slightly more complex and demanding PIC projects. The book has a supporting web site, www.wileyeurope.com/college/parchizadeh, containing programs from the book and most of the schematic diagrams, strip board and PCB layouts for the circuits used in individual projects.

The book is divided into six chapters. The first chapter introduces the basic characteristics of PIC MCU using one particular member of the family – PIC16F627A. Basics of C and assembler programming for this chip are discussed next. Finally, a development cycle for a simple program is explained. The MPLAB Integrated Development Environment (IDE) used to develop and test all of the programs in this book is introduced in the second part of the chapter. Some specific debugging features and simulation capabilities of MPLAB and MPSIM debugging tool are discussed at the end of the chapter.

Chapter 2 introduces simple circuits used to interface PIC to LEDs, push buttons and seven-segment type displays. Programs to switch the LED on and off, read the state of the push buttons or display a single digit on a seven-segment LED display are all developed in this chapter and discussed in some detail. Circuits developed in this chapter are based around PIC12F629 and PIC16F627A microcontrollers.

In Chapter 3, two more display circuits based around the PIC16F627A microcontroller are developed and their operation is described in detail. First, a four-digit, seven-segment, LED display PIC interface is designed and two programs using C and assembler are developed to make use of this circuit. Programs count from 0 to 9999 and display the current count on the four-digit seven-segment LED. In the second part of the chapter the design interface and operation of an LCD display circuit is discussed and a program to display a simple message on the LCD is explained.

Chapter 4 deals with an RS232 interface. A circuit to make use of this interface and enable PIC to communicate with PC is designed and discussed. Programs to send or receive a set of characters from PIC to PC and vice versa are given and described.

Chapter 5 introduces a more advanced member of PIC family, PIC18F252, and some important features of this chip. The first program in this chapter is developed to give the reader a clear understanding of the use of the analog-to-digital converter (ADC) available on this chip. Using built-in ADC, an analog voltage supplied to one of the PIC pins is converted to digital form; its value is constantly checked in the program and compared to some preset value defined at the beginning of the program. The use of an internal timer, also available on PIC MCUs, is described next. Timers can be used to achieve accurate timing, which is often required in various embedded type systems containing PIC micro-controllers. This chapter is completed by an explanation of the way in which an advanced timer interrupt feature can be used to make PIC programs more efficient and versatile. Other interrupt sources, available on most PIC MCUs, are not discussed in this book but readers should be aware that this topic is by no means exhausted with the two simple programs given at the end of Chapter 5.

Finally, Chapter 6 contains some more PIC projects and circuits based around PIC12F675, PIC16F627A and PIC18F252 models, which should give the reader an indication of the versatility of PIC microcontrollers and systems based around those devices. They should also give the reader good, clear ideas for further work and investigation in this field.

Acknowledgements

The authors would like to take the opportunity to thank all those individuals and/or companies that have contributed or helped in some way in the preparation of this book. Our thanks go to:

- Department of Electronic and Computer Engineering, University of Portsmouth, for providing us with the time and facilities to write this book.
- Technical Staff of the Department of Electronic and Computer Engineering, University of Portsmouth, Steve Webb, Andrew Farrar and Peter Richmond for their continuous support.
- Dr Dragana Nikolic for reading and correcting big parts of this book.
- Microchip Semiconductors, HI-TECH Software and CadSoft Computer, Inc., for providing free versions of MPLAB, C18, Hi-tech Lite, and EAGLE software for development and implementation of the software and hardware of the projects in the book.

Special thanks go to our families for their love and support: Hassan's wife Hoory, daughters Nadia and Layla and son Nima and Branislav's wife Jasmina and daughters Zoe and Sophia. Finally, Hassan would like to dedicate his contribution to the memory of his mother Baygam and father Jaffar, whom he misses dearly.

1

Preparing to do a PIC Project

1.1 Introduction

The aim of this chapter is to consider a number of issues that need to be taken into account before doing almost any microcontroller-based project. First, the reader will be introduced to a PIC (programmable interface controller) microcontroller by a brief discussion of one of the models from the PIC microcontroller family – the PIC16F627A. This model is now a common choice for low-cost PIC projects and has practically replaced the very popular PIC16F84 model. The PIC16F627A is therefore a choice for a large number of projects from this book although some other simpler and more complex models are also being used. The rest of the PIC family will be considered briefly, introducing some other models used for the projects in this book. We will then discuss the basics of two programming languages commonly used to develop PIC programs in practice and throughout this book – assembly and C. This will by no means be a detailed discussion of those two languages; a separate book would be needed for that. The aim instead is to provide a short overview of the basic features of both languages and to enable readers to learn the rest of it while doing projects from the other chapters of this book. Material covered in this chapter should therefore be sufficient to allow the reader to start with the first programs and projects from Chapter 2 and gradually build knowledge to do more complex projects from the rest of the book. Finally, we will demonstrate how to develop and test a simple PIC program using the MPLAB® -Integrated Development Environment (IDE).

1.2   Overview of PIC Microcontroller

The name PIC denotes several families of microcontrollers manufactured by Microchip Technology. This range is huge and very versatile so discussing even a small number of microcontrollers would be a difficult and time-consuming task. Instead, in this section, we will concentrate on the basic features and layout of one of the most popular members of the mid-range PIC16 family: PIC16F627A. Once the basic features of this device are explained it will be easier to introduce and understand the operations of more complex PICs used in later chapters of this book.

1.2.1   PIC16F627 Building Blocks

Every computer system, however complicated or simple, consists of a number of common building blocks. Those are: the CPU (central processing unit or microprocessor) block, the memory block (RAM and ROM) and the input/output (I/O) block (interface circuitry). The CPU performs all the logic and arithmetic functions; memory is used to store programs and data while the interface provides means of communication and data exchange between the microcomputer system and the external world.

A microcontroller is a stripped-down version of the computer system architecture with one important difference – all of the system blocks are placed on one chip. The microcontroller-based system therefore requires very little additional circuitry for its proper operation. All that is needed in most cases is a clock input to provide timing for the system operation.

The PIC16F27A microcontroller contains all of the previously mentioned blocks. Components of this microcontroller, described in slightly more detail, are:

- *The CPU.* The 'brain' of a microcontroller. It is responsible for finding and fetching the right instruction to be executed, for decoding that instruction, and finally for its execution.
- *Memory.* Split into two physically separate blocks – program and data memory. This so-called Harvard architecture is used to speed up the operation of the microcontroller as both data and instructions can be fetched from separate memories using separate buses simultaneously.

- *Program memory*. Used to store a program to be executed in the central processing unit of the microcontroller. It is of flash type so the microcontroller can be programmed many times before a system developer is happy with its performance and the programmed PIC is finally installed into some bigger system. If the power to the micro-controller is switched off, the content of the flash-type memory is not lost. The size of the program memory on the PIC16F27A is 1024 words (1 kwords), where one word holds 14 bits.
- *Data memory*. Used to store microcontroller data. It is further divided into EEPROM and RAM memory: EEPROM memory holds important data that need to be saved when there is no power supply to the microcontroller; RAM is used by a program to store inter-results or temporary data during the program execution. EEPROM contains 128 bytes of data whereas RAM holds 224 bytes (1 byte contains 8 bits so the widths of program and data memories are different).
- *PORTA and PORTB*. Physical connections between the microcon-troller and the outside world. Both of those ports have eight pins and those pins are bidirectional – they can be used for input or output of data provided they are properly configured as input or output pins in the program. The exception is pin 5 of port A (RA5), which is an input-only type pin. Two special function registers within PIC, TRISA and TRISB, control the direction of the port pins. Writing '1' in the particular bit of the TRISA register configures the corresponding pin of port A as an input pin; '0' in TRISA makes it an output pin. The same is true for the port B pins and TRISB register. Some of these port pins are multiple-purpose pins and can be used for other periph-eral functions of the processor. This will be explained in more detail in Section 1.2.3 where the layout (pin out) of the PIC16F27A chip is discussed.

The PIC blocks mentioned above communicate through a complex system of communication lines called *buses*. The *data buses* are used for the transfer of data through the system and *address buses* communicate addresses of data and program instructions to be accessed during pro-gram execution. Various other communication lines exist in the PIC and those are usually referred to as *control bus* lines. Note that since two separate memories exist in the PIC, both data and address bus systems are doubled, i.e. PIC16F27A (Figure 1.1) has a data memory (DM) address bus as well as program memory (PM) address bus. Similarly, this processor also has a DM data bus and PM data bus.

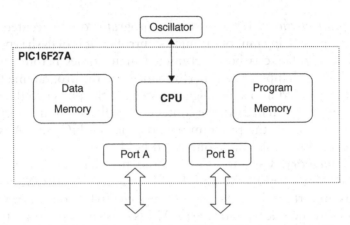

Figure 1.1 Main elements of the PIC16F27A microcontroller

1.2.2 EEPROM and RAM Memories on PIC16F27A

The EEPROM memory on the PIC16F27A holds 128 bytes of non-volatile information. This memory is electronically programmable so it is not a fast RAM-type memory and it can be awkward to access it within the program. It is normally used to store data that is not frequently changed.

The RAM memory on the PIC16F27A is actually split into four memory banks where each bank holds access to 80 memory locations. This, however, does not mean that the total capacity of the RAM memory on the PIC16F27A is 320 bytes. It is more complicated than that. Here is the explanation.

This memory can be considered to consist of two different types of registers – special function registers (SFRs) and general purpose registers (GPRs). The first 32 bytes of each memory bank (00h-1Fh) belong to SFRs. Those registers are used by the CPU to control the desired operation of the device and to record the operating states of the PIC, the I/O port conditions and the other conditions. Not all of those registers are implemented on the PIC16F27A model. There are 21 SFR bytes in bank 0, 18 in bank 1 and seven SFRs in banks 2 and 3. Some of these registers will be used and described in more detail in the remaining chapters of this book.

The GPRs are placed in the first three memory banks – 80 bytes of GPR in banks 0 and 1 and 48 in bank 2. Those registers can be used to store results and conditions temporarily while the program is running. The content of the GPR is lost when the power is switched off. The rest of the available memory space is not used – if it gets accessed, it reads as 0. This memory arrangement is shown schematically in Figure 1.2.

	Bank 0		Bank 1		Bank 2		Bank 3
00h		80h		100h		180h	
01h	TMR0	81h	OPTION	101h	TMR0	181h	OPTION
02h	PCL	82h	PCL	102h	PCL	182h	PCL
03h	STATUS	83h	STATUS	103h	STATUS	183h	STATUS
04h	FSR	84h	FSR	104h	FSR	184h	FSR
05h	PORTA	85h	TRISA	105h		185h	
06h	PORTB	86h	TRISB	106h	PORTB	186h	TRISB
07h		87h		107h		187h	
08h		88h		108h		188h	
09h		89h		109h		189h	
0Ah	PCLATH	8Ah	PCLATH	10Ah	PCLATH	18Ah	PCLATH
0Bh	INTCON	8Bh	INTCON	10Bh	INTCON	18Bh	INTCON
0Ch	PIR1	8Ch	PIE1	10Ch		18Ch	
0Dh		8Dh		10Dh		18Dh	
0Eh	TMR1L	8Eh	PCON	10Eh		18Eh	
0Fh	TMR1H	8Fh		10Fh		18Fh	
10h	T1CON	90h					
11h	TMR2	91h					
12h	T2CON	92h	PR2				

Figure 1.2 Memory map of the PIC16F27A microcontroller

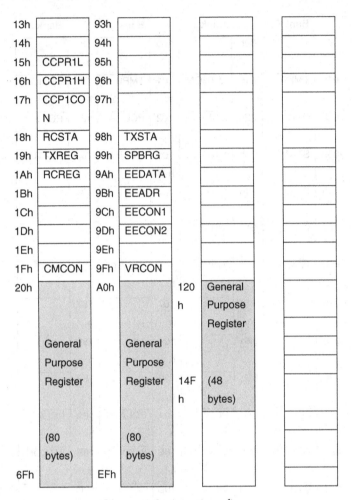

13h		93h				
14h		94h				
15h	CCPR1L	95h				
16h	CCPR1H	96h				
17h	CCP1CON	97h				
18h	RCSTA	98h	TXSTA			
19h	TXREG	99h	SPBRG			
1Ah	RCREG	9Ah	EEDATA			
1Bh		9Bh	EEADR			
1Ch		9Ch	EECON1			
1Dh		9Dh	EECON2			
1Eh		9Eh				
1Fh	CMCON	9Fh	VRCON			
20h	General Purpose Register (80 bytes)	A0h	General Purpose Register (80 bytes)	120h	General Purpose Register (48 bytes)	
6Fh		EFh		14Fh		

Figure 1.2   (*continued*)

1.2.3 PIC16F27A Pins

We will complete this section with a short description of all 18 pins of the PIC16F27A microcontroller shown in Figure 1.3.

- RA0 to RA7 are the eight pins of port A. Port A is a bidirectional port, which means it can be configured as an input or an output. The number following RA is the bit number (0 to 7). So, we have one 8-bit directional port where each bit (with the exception of bit 5) can be configured as input or output. As shown in Figure 1.3, all of the port-A pins have alternative functions.

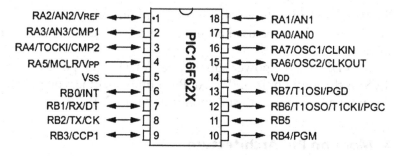

Figure 1.3 Pin-out diagram of the PIC16F27A microcontroller

© Microchip Technology Inc. Reproduced with permission.

- Pins 1, 2, 17 and 18 can be configured and used as analog inputs (AN0 to AN3). Those pins are therefore attached to internal comparators of the PIC. Voltage in the range 0–5 V on those pins is converted into digital form and further processed by the microcontroller.
- Pin 4 can be used as a master clear – reset pin (MCLR). Reset is used for putting the microcontroller into a 'known' (default) condition. Upon setting this pin to 0 V, all of its registers will be placed in a starting position. Here we use internal reset circuitry activated when the processor is powered up (power-up reset). This option is normally used when the microcontroller does not behave in an appropriate way due to some undesirable condition.
- Pins 16 and 17 (OSC1/CLKIN and OSC2/CLKOUT) can be used when an external oscillator or crystal/RC timing elements are used to provide timing for the microcontroller.
- Pin 3 (TOCK1) can be used as an input for the Timer 1 module. It operates independently from the main clock.
- RB0 to RB7 are the eight I/O pins of port B. Port B is a second bidirectional port and it behaves in almost the same way as port A. Alternative functions of port pins are different for port A and port B.
- Pin 6 can be configured and used as an external interrupt pin to detect external events. An event is detected when the interrupt pin changes state from '1' to '0' or from '0' to '1' (programmable).
- Pins 7 and 8 can be used for serial communications: TX is the asynchronous serial transmit pin – data is sent from the chip on this pin; RX is the serial receive pin and data is sent to the chip on this pin.
- Pin 9 can be used as a capture/compare pin (CCP1) in order to measure duration of external events (the length of the PWM – pulse width modulated pulse).

- Pin 10 is used for a low-voltage programming of the PIC.
- Pins 12 and 13 can be used as an oscillator and as timer inputs for timer 1.
- VSS and VDD are the power supply pins. VDD is the positive supply and VSS is the low supply or 0 V. All PIC family members operate off a 5 V supply and some can use supplies down to 2 V.

1.2.4 More on PIC Architecture

Trying to explain the architecture of PIC microcontrollers in more detail would probably take another book or at least another long and not-very-interesting chapter of this book. Instead, we will provide a very brief insight into some of the PIC features useful for understanding and doing some interesting PIC projects explained in the other chapters.

- *Timing*. An oscillator (internal or external) is generally used to drive the PIC by clocking data and instructions into the processor. The actions of the CPU are caused by every fourth oscillator pulse, which makes the instruction times easy to calculate. Most instructions take one clock cycle (four oscillator pulses), so with a 4 MHz oscillator it will take $1\,\mu$s to execute each of those one-cycle instructions.
- *Program execution control*. The program counter is an internal 13-bit CPU register used to store the current program position. After each loaded instruction the program counter is incremented automatically so that it points to the location of the next instruction in the program memory.
- *CPU status*. Bits of the status register contain the information about the status of the arithmetic and logic unit from the CPU. Those bits are updated after certain instructions that modify the main working register content of the CPU. The PIC16F27A has an 8-bit status register, which contains information about the memory bank currently being accessed, carry, power state of the PIC, borrow or zero results of the executed instruction.
- *Timers*. To provide accurate timing for the microcontroller actions a special function register called a timer can be used. This register is connected to the internal clock and increments at the clock frequency divided by four. When it rolls over from its maximum count (255 for an 8-bit timer, 65535 for a 16-bit timer) to zero, a flag is set to signal that event. It would not take long to count from 0 to the maximum count at 1 MHz, so a programmable prescaler can be used in combination with the timer module. The prescaler can be set to give out

pulses at ratios of 1:2, 1:4 and so forth, up to 1:256, extending the timeout up to tens of milliseconds range for a 4 MHz oscillator used in the system.

• *Interrupts*. An improved solution to exact timing described above is to set up a timer to generate an interrupt. A routine can be set by the programmer to be executed automatically every time the timer overflows (rolls over at the maximum count). This requires some programming skill and will be explained in more detail in Chapter 5. There might be other situations that can cause an interrupt routine to be executed, such as a change of state on the port B pins or some external event sensed by the interrupt pin RB0.

• *Watchdog timer*. To provide a means of recovery from some system problems a watchdog timer is implemented on all PIC microcontrollers. During its execution, the program needs to reset the watchdog timer at predetermined intervals, but if it fails to do so, due to a problem in the execution, the watchdog timer will initiate the reset itself. This can be a useful option in case the program goes into an endless loop or some hardware problem occurs to prevent the program operating correctly.

1.2.5 Brief Overview of the PIC Family

The whole PIC family can be divided in a number of distinct groups:

• *Baseline core devices*, represented by the PIC10 series, as well as some PIC12 and PIC16 devices. Baseline devices are available in six-pin to 40-pin packages.
• *Midrange core devices*, labelled PIC12 and PIC16.
• *PIC17 and PIC18 high-end core devices*, represented by a not-very-popular PIC17 series, produced in packages from 40 to 68 pins and later superseded by the PIC18 architecture.
• *PIC24 and dsPIC microcontrollers* are 16-bit microcontrollers. The PIC24 devices are designed as general-purpose microcontrollers and dsPICs include digital signal processing capabilities.
• *PIC32MX* – these 32-bit microcontrollers are the latest addition to the PIC family introduced in November 2007.

1.3 Basics of PIC Assembly Language

Assembly languages are closer than high-level languages to possessing a one-to-one correspondence between symbolic instructions and executable machine codes. They are also usually more difficult to use. The aim of this section is to give a brief introduction to some aspects of PIC programming using assembler language. Most of the microcontroller programs spend a lot of time on the actual movement of data through the device. This is certainly the case for assembler language-type programs so we will start our introduction to PIC assembler language by looking into some instructions that are commonly used to move data through the microcontroller.

1.3.1   Data Movement, Arithmetic and Logical Operations

The PIC16F27A has a very small set of instructions – there are only 37 of them. Since the width of the program memory is 14 bits and the address of each register can take up to eight bits, two registers cannot be used in the same instruction. Remember that we also need to specify what action we actually want to accomplish with each particular instruction – too much information to fit into just 14 bits!

To move data from one register to the other it is therefore necessary to use intermediate register storage. The working register, labelled W, is used to accomplish this task. This is an internal CPU register that does not have a specific address. Thus, movement of data from one register to the other will require two assembly language instructions.

Suppose we want to move the value of variable MYDATA to PORTB in order to output it to some external device connected to PORTB. The following instruction sequence will accomplish this:

```
1 movf      MYDATA,W;  copy MYDATA into working register W
2 movwf     PORTB   ;  copy working register W into PORTB
```

The first instruction is of the form `movf f,d`, which moves the register or memory location f to the destination specified with d. In the above sequence d is specified as our working register W so the value of MYDATA is copied to the working register W. Other possibilities for d are 1 and 0. If d is 0, the result is the same – f is copied over to register W. If d is 1, the MYDATA register will be copied to itself. This option will have no effect on the content of the register because we are effectively overwriting the register f with the same value. It does however effect the Z flag in the PIC status register and can therefore be a useful option in some situations. We will not discuss this effect further in this section.

The second instruction is of the form `movwf f` and it simply moves whatever is in W into the register f. Variable MYDATA remains unchanged in the first instruction and W remains unchanged in the second. So it is useful to keep in mind that it is a copy rather than a move action we have achieved through this sequence.

Loading a register with a literal (8-bit value) also takes two instructions and involves the use of working register W. Here is an example of loading variable MYDATA with the literal binary number 11110000 (hexadecimal F0, or F0h). This effectively clears lower four bits of the working register W and sets the upper four bits of the same register:

```
1 movlw 0xF0          ; put the number/literal F0h into W
2 movf MYDATA                        ; copy W into MYDATA
```

To summarize here:

- `movf f,d` copies the content of specified register (f) to working register W or to itself, depending on the destination specification d (W, 0 or 1)
- `movwf f` copies the content of the working register W to specified register f
- `movlw l` copies/moves specified 8-bit literal l to the working register W.

A common mistake by novice PIC programmers is to confuse loading some register with a value from W and loading the working register W with a value from some other register. Care needs to be taken when using the data movement instructions listed above.

A similar approach must be taken when, instead of copying, we want to apply arithmetic or logical functions (for example, addition, subtraction, logical AND, OR, XOR and others):

```
1 movlw k                             ; move number k into W
2 subwf f,d ; subtract W from f and put the result according to d
```

The result of the first line of the above sequence should already be clear to us. Literal k is copied to working register W. The second line subtracts W from f and stores the result into the destination specified by d ; d can again be specified using one of three options – W, 0 and 1. In case d is specified as W or 0, the result is stored back to W, f is not changed and, if d is 0, the result is stored in f and W is unaffected. Of course, in the above sequence the value k needs to be replaced with some real value if the program is to work properly (such as 0xF0).

Table 1.1   Common single-operand arithmetic and logic instructions
(W is source and destination register)

Syntax	Description
addlw k	put the value of 'k + W' in W
sublw k	put the value of 'k − W' in W
andlw k	put the value of 'k and W' (logical 'and' operation) in W
iorlw k	put the value of 'k ior W' (logical 'inclusive or' operation) in W
xorlw k	put the value of 'k xor W' (logical 'exclusive or' operation) in W

Single instructions to accomplish some arithmetic or logical operation will work if the value to be changed is already in W and the result destination is again W. Examples are given in Table 1.1.

PIC assembly language also has some instructions where there is no choice between the number of operands to be used in the instruction – only one operand can be used. Commonly used single-operand instructions are given in Table 1.2.

We have seen that the move instruction of type movwf f does not have a choice of destination – it is always the register f specified in the instruction. PIC has more 'WF'-type instructions but the rest of them have a choice of destination – it can be the working register W or the other register f, depending on the destination specified. These instructions are listed in Table 1.3.

Some other useful instructions are listed in Table 1.4. It is important to remember that the result of those instructions can be stored in the working register W (when d is specified as W or 0) or in the specified file register f (when d is specified as 1).

1.3.2 Program Flow Control

The instructions listed and discussed above will probably be the ones you will need to use in order to perform some simple arithmetic and logical

Table 1.2   Exclusive single-operand instructions

Syntax	Description
clrf f	set all bits in specified register f to zero (clear register f)
clrw	set all bits in register W to zero (clear register W)
bcf f,b	set bit b in register f to zero (bit clear bit b in f)
bsf f,b	set bit b in register f to one (bit set bit b in f)

Table 1.3 WF-type instructions

Syntax	Description
addwf f, d	put the value of 'f + W' in destination d (either W or f)
subwf f, d	put the value of 'f − W' in destination d (either W or f)
andwf f, d	put the value of 'f and W' (logical 'and' operation) in destination d (either W or f)
Iorwf f, d	put the value of 'f ior W' (logical 'inclusive or' operation) in destination d (either W or f)
xorwf f, d	put the value of 'f xor W' (logical 'exclusive or' operation) in destination d (either W or f)

Table 1.4 Common two-operand type instructions

Syntax	Description
incf f, d	put the value of 'f + 1' in destination (specified with) d
decf f, d	put the value of 'f − 1' in destination d
comf f, d	put the result of toggling all bits from f in d
swapf f, d	put the result of swapping the nibbles in f in destination d
rlf f, d	put the result of rotating f left through carry in destination d
rrf f, d	put the result of rotating f right through carry in destination d

data operations or to move data between the various destinations in the microcontroller. To make your programs really useful and more complex you also need to be able to control the flow of the program. Normal sequential execution of the program assumes that the program starts with the first instruction, then it goes on to the next one in the list and continues in a similar manner. The instruction to disrupt this kind of operation would need to involve an abrupt change in the program counter content. Two instructions from the PIC set can do this – goto and call. However, those instructions cannot be used interchangeably as there is one important difference in the way they operate. The goto instruction will simply make the program execute the instruction at the specified location in the program memory; for example, goto 0x123 will set the program counter to the value of 123h. This, in turn, will cause the instruction at location 123h to be executed by the CPU next. Instructions following the one at location 123h would be executed and the program would not return to where it left off before being made to execute the instruction at location 123h. However, call 0x123 would push the address of the next instruction in the program on the stack (dedicated memory space for exactly this kind of situations) and then set the program counter to 123h, thus making the program execute the instruction at that location next. Instructions at the locations following the location 123h will continue to be executed sequentially but only until the return instruction is encountered. This instruction will cause the retrieval of the address most recently stored on the stack into program counter. The program will therefore return to the instruction following the `call 123` instruction and continue execution from there. This is normally used when writing the subroutine part of the program. The main program is made to call the subroutine to execute a specific sequence of instructions and after the execution of the subroutine the program needs to return to where it left off before the call to the subroutine. It is possible to nest a number of subroutines within each other: one subroutine can call another subroutine, which in turn can call another subroutine and so forth. The number of nested subroutines is determined by the size of the stack. In the PIC16F27A, the stack consists of eight words used in circular manner: after the eighth word, it rolls over to the first. It is therefore possible to nest eight subroutine calls. If we try to nest more than eight subroutines the program will become lost while trying to return from the subroutine to a proper instruction in the calling subroutine or the main program. The program cannot find its way back and will act in a strange and unexpected way. The program will continue to execute but the expected results will not be there. This is a stack overflow problem, which is difficult to detect when testing the program.

There are two more versions of the return instruction for PIC16F27A. All three return instructions are listed in Table 1.5.

Another way to control the flow of the program implemented on the PIC microcontroller and a number of other microprocessors is simply to skip the next instructions depending on the result of the current operation. Those instructions are listed in Table 1.6.

1.3.3 Example Assembler Program

We are now ready to write and understand our first PIC assembler language program. It is shown in Listing 1.1.

Table 1.5 Return from subroutine instructions

Syntax	Description
`return`	return from call, location of the next instruction is retrieved from the stack and copied into the program counter
`retlw k`	as 'return' but literal k is also placed into W
`retfie`	return from interrupt

Table 1.6 'Skip-the-next-instruction' type instructions

Syntax	Description
incfsz f,d	put the value of 'f + 1' in destination d (either W or f) and skip the next instruction if the result of this increment is zero
decfsz f,d	put the value of 'f − 1' in destination d (either W or f) and skip the next instruction if the result of this decrement is zero
btfsc f,b	test bit b (but do not change it) of register f and skip the next instruction if the bit is clear – bit test skip clear
btfss f,b	test bit b (but do not change it) of register f and skip the next instruction if the bit is 1 – bit test skip clear

```
 1 ;********************************************************
 2 ; this program outputs two binary patterns on port B   *
 3 ; one after the other – 01010101 and 10101010           *
 4 ;********************************************************
 5
 6 include "p16F627A.inc"
 7
 8 bsf        STATUS,RP0   ; set RP0 bit = select bank 1
 9 movlw      0x00
10 movwf      TRISB        ; set port B pins to all outputs
11 bcf        STATUS,RP0   ; clear RP0 bit = select bank 0
12
13 loop       movlw 0x55
14            movwf PORTB  ; PORTB = 0x55 = 0b01010101
15            movlw 0xAA
16            movwf PORTB  ; PORTB = 0xAA = 0b10101010
17            goto loop
18            end
```

Listing 1.1 First assembler program

The first line of the program (line 6 in Listing 1.1) contains the include statement, which tells the assembler program to include the specified 'inc' type file into assembly procedure. While assembling the binary version of our program, assembler looks in the quoted file (p16F627A.inc) for any symbols not defined within the program. The p16F627A.inc file holds definitions of all SFRs in the PIC together with their addresses. Each device in the PIC family will have its own .inc file with definitions specific for that particular device, so care needs to be taken to include the proper file.

As long as there is a power supply to the microcontroller, this program outputs two-bit patterns on eight pins of port B – 01010101 (55h) and 10101010 (AAh). The first bit pattern is moved from the working register W to port B in line 14 of the program and the second bit pattern is output to port B in line 16. Previously those two patterns were loaded into the working register in lines 13 and 15. Line 17 redirects the program execution back to line 13 and this process is repeated indefinitely.

The first part of the program (lines 8–11) is used to set port B as an output port. This is done by setting corresponding bits (all of them in this case) of register TRISB low. This is done in lines 9 and 10 of the program where 00h is first moved to working register W and then from W to TRISB register. This register is located in Bank 1 as shown in Figure 1.2 and the default active Bank on PIC is Bank 0. To switch to Bank 1 the RP0 bit of the STATUS register needs to be set to 1. This is done in line 8 of the program. Bank 0 is again activated at line 11 of the program by resetting RP0 bit back to 0.

Descriptions of each program line following the semicolon are just program comments. Everything following the semicolon sign is ignored by assembler and is only used by the programmer to supply some useful documentation to the code. This can be extremely helpful when trying to understand a program that somebody else has written or even a program that you wrote some time ago.

After typing and saving this code we would need to invoke PIC assembler to assemble the file containing the code – to translate it into machine code. Machine code is a pattern of 0s and 1s understandable and executable by the CPU. We would then need to test the program and run it on the PIC16F27A microcontroller. This process will be explained in the last section of this chapter where the MPLAB development environment is considered. The next section will discuss the development of PIC programs using the C programming language.

1.4 Introduction to C Programming for PIC Microcontroller

Sometimes, when more complex algorithms need to be implemented, assembler language programming can lead to cryptic and difficult-to-understand programs. A lot of programmers nowadays tend to use high-level languages, rather than assembler, to program microcontrollers. Although programming in high-level language is generally much easier, there are also disadvantages. Programs written in high-level language are larger so more program memory might be needed to store them. They can also be less efficient when it comes to speed of execution – a high-level language program is typically slower compared to a similar program written in assembler language. The high-level language most frequently used for microcontroller programming is C. One of the advantages of using C compared to some other high-level languages is that various C compilers (programs that translate C programs into machine code) are usually very efficient. A lot of manufacturers claim their compiler to be 80% efficient. This basically means that the corresponding assembler language program will only be 20% smaller in size on average compared to our program written in C. This tradeoff between the ease of programming and the program size is usually acceptable in most situations, although you might find that this figure of 80% efficiency is difficult to achieve in practice. C language efficiency is usually somewhere in the 50%–70% region.

This section will introduce some basics of C programming for PIC microcontrollers. Even though C is a standard programming language, there are some specific issues to consider when programming any microcontroller in C. For a more general and detailed treatment of C language, the reader is advised to look at some standard C books, while the main aim of this section remains to introduce the basics of C language. Enough information will be given to start with some simple PIC projects explained in Chapter 2 and then build your knowledge further as you attempt more and more complex PIC projects.

1.4.1 Template C Program for PIC16F27A

C is a function-based language and, in fact, any C program is merely a function. Functions are sections of code that perform a single, well-defined task. Functions are the C equivalent of assembly-level sub-routines. A function in C can take any number of parameters and return a maximum of one value. A parameter (or argument) is a value passed to a function that is used to alter its operation or indicate the extent of its operation. A function consists of a name followed by the parentheses enclosing arguments, or a keyword 'void' if the function requires no arguments. If there are several arguments to be passed, the arguments are separated by commas. A simple C function is detailed in Figure 1.4.

Every C program must at least have a function named 'main', and this function is the one that executes first when the program is run. Here we can write the simplest possible version of the main function that can run

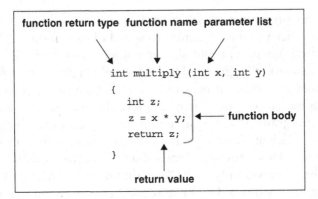

Figure 1.4 Simple C function

```
1  // template C program – does nothing useful – starts and
   stops
2  #include <pic16f2xa.h>
3  void main(void)
4  {
5
6  // body of the program goes here
7
8  }
```

Listing 1.2 Template C program for PIC microcontroller

on the PIC16F27A microcontroller. The program given in Listing 1.2 will do almost nothing – it finishes as soon as it starts. However, it represents a useful template for further development of more complex C programs on our PIC microcontroller.

The second line in Listing 1.2 is the so-called 'include' statement. As in assembler language this line tells the C compiler that during the compilation it should look into header file (pic16f2xa.h) for any symbols not defined within the program. This file holds definitions of all SFRs in the PIC and their addresses. Different models of PIC microcontroller usually require different .h header files with definitions specific for that particular device. Various C compiler manufacturers usually provide their own versions of each header file, specific to and compatible with their own compiler. Care therefore needs to be taken to include a proper .h file in your code.

Line 3 from Listing 1.2 declares the main function. The prefix void in this line means that this function returns no argument and the second void means that the function also requires no arguments to be supplied to it. This is almost always the case with PIC C programs, so you will use this construction very often when programming PIC using C language. Line 4 announces the beginning of the main function and line 8 is the end of the function. C functions are always enclosed in curly brackets and the main function is no exception. The fact that there is no real code between those two brackets only means that this C program will do nothing useful – it will start and soon after that it will end. Lines 1 and 6 are program comments.

Everything starting with a double forward slash is a program comment. Like comments in assembler language programs, comments in C programs are ignored by the C compiler and are used to clarify some sections of the code to whoever needs to read, use or modify the program.

The C program in Listing 1.3 is slightly more meaningful.

```
1  // this program sets all 8 pins of port B to high (logic 1)
2  #include <pic1684.h>
3  void main(void)
4  {
5
6      TRISB = 0x00;
7      PORTB = 0xFF;
8
9  }
```

Listing 1.3   Simple C program

This program will set all eight pins of port B to logic 1. Note that the body of the main program consists of just two statements. The statement in line 6 configures all pins of PORTB as outputs by writing logic 0s into the TRISB register and the statement on line 7 sends the number 0xFF to port B causing all pins of that port to go high (logic 1 state). The common way to terminate statements in C code is to use a semicolon at the end of each line, as has been done in the statements in lines 6 and 7 (there are some exceptions not discussed here).

To complete this brief review of C functions we will rewrite the program from Listing 1.3 by introducing a new function called myfunc. This function will have a task of sending the number 0xFF to port B. This is, of course, not a clever thing to do – the program itself is so short that there is no point in splitting it even further. We just do it here in order to demonstrate how a C function can be called by the main program, or how to call a function from any other function in a C language program.

```
 1  // this program sets all 8 pins of port B to logic level 1
 2  // it calls a function myfunc to accomplish this task
 3  #include <pic1684.h>
 4  void main(void)
 5  {
 6      TRISB = 0x00;
 7      myfunc();
 8  }
 9
10  void myfunc(void)
11  {
12      PORTB = 0xFF;
13  }
```

Listing 1.4 Calling another function from the main function in C

In the program from Listing 1.4, the main function contains two statements. The first statement (line 6) configures PORTB as output and the second statement is a call to the function, myfunc, which contains a further, single statement. It copies the hex value of 0xFF to the PORTB register, thus effectively setting all pins of port B to a logic high state.

1.4.2 Variables

Variables in C are data objects that may change in value during the program execution. Variables must be declared immediately after the curly bracket marking the beginning of a function by specifying the name and data type of the variable. Variable names in C may consist of a single letter or a combination of a number of letters and numbers. Normally up to 52 characters can be used for the variable name. Starting a variable name with a number is not permitted in C and spaces and punctuation are not allowed as part of a variable name. C is a case-sensitive language and two variables 'A' and 'a' are therefore treated as two separate and different variables. In C, a data type defines the way the program stores information in memory. Data types with corresponding specifiers (proper ways of declaring data types in C), allocation sizes and ranges for C18 compiler available from Microchip and designed specifically for PIC18 devices are given in Table 1.7. This is not a universal table as it might change depending on the C compiler.

Defining the type of the variables in a PIC C program is an important factor in the efficiency of a program. The compiler will reserve a memory space sufficient to save a variable according to its type. Larger data types will usually mean longer computation time so if speed is more important than range you should try to make most of the variables in your program

Table 1.7   Basic C data types

INTEGER DATA TYPES

Type	Size	Minimum	Maximum
char[1,2]	8 bits	−128	127
signed char	8 bits	−128	127
unsigned char	8 bits	0	255
int	16 bits	−32768	32767
unsigned int	16 bits	0	65535
short	16 bits	−32768	32767
unsigned short	16 bits	0	65535
short long	24 bits	−8,388,608	8,388,607
unsigned short long	24 bits	0	16,777,215
long	32 bits	−2,147,483,648	2,147,483,647
unsigned long	32 bits	0	4,294,967,295

FLOAT DATA TYPES

Type	Size	Minimum Exponent	Maximum Exponent	Minimum Normalized	Maximum Normalized
float	32 bits	−126	128	2^{-126} $\approx 1.17549435e-38$	$2^{128} * \left(2 - 2^{-15}\right)$ $\approx 6.80564693e+38$
double	32 bits	−126	128	2^{-126} $\approx 1.17549435e-38$	$2^{128} * \left(2 - 2^{-15}\right)$ $\approx 6.80564693e+38$

© Microchip. Reproduced with permission.

the 'char' type. While doing this you need to be careful and keep in mind the range of the 'char'-type variables. If, for example, the 'unsigned char' variable exceeds the 0–255 range your program will give incorrect results. Since most of the registers on PIC16F27A are 8-bit registers, 'char' is usually sufficient for most variables in the program. To give you an idea how to declare your data in C programs some examples are given below:

```
1 int goals;       // integer variable
2 float x,y,z;     // three float variables
3 char p = 0xFF;   // char variable is declared and
4                  // initialised at the same time
5 double f = 56.3; // double variable
```

1.4.3 Arrays

Arrays in C are specific data structures used to store multiple variables of the same data type. An array of 10 character variables named A can be defined as:

```
char A[10];
```

The above declaration actually declares the array with 10 elements: A[0], A[1], A[2], . . . , A[9], where the value within brackets is called a subscript. In the C language the array subscript starts from 0, so to access all elements of the array A, the subscript needs to take values from 0 to 9. Accessing A[10] is not legal and this can be a source of errors in many programs!

If we want to assign the initial values to elements of an array, we need to enter these values between curly brackets:

```
char days[7] = {1, 2, 3, 4, 5, 6, 7};
```

To access any value from the array days, we need to use [ ] separators.

```
Tuesday = days[1];    // second value from the array days is
                      // assigned to a variable named Tuesday
Sunday = days[6];     // seventh value from the array days is
                      // assigned to variable named Sunday
Monday = days[7];     // error!!!, days[7] does not exist
                      // some random value will be assigned
                      // to a variable named Monday
```

1.4.4 Constants

C also allows declaration of constants. When you declare a constant, it is rather like a variable declaration except the value cannot be changed later in the program. The attribute 'const' is used to declare constant as shown below:

```
int const a = 1;
const int b = 4;
```

1.4.5 C Operators

C has a full set of arithmetic, relational and logical operators. It also has some useful operators for direct bit-level manipulations on data, which makes it similar to assembler language. Most important operators are listed in Table 1.8. These operators can be used to form expressions. Mathematical operators follow standard precedence rules – multiplication and division will be executed before any addition or subtraction. Sub-expressions in parentheses have the highest precedence and are always evaluated first. Statements may optionally be grouped inside pairs of curly braces { } and as such are called compound statements.

A common mistake in writing the C statements is to use assignment operator '=' instead of equality operator '=='. 'i = j' and 'i == j' are both perfectly legal C statements but the first one will copy the value of variable j into i while the second compares the values of two statements and results in a logical value ('1' if those two variables are equal, '0' if i and j are different).

Table 1.8   Common C operators

	Operator	Action	Example
Assignment operators	=	Assignment	x = y;
Mathematical operators	+	Addition	x = x + y;
	−	Subtraction	x = x − y;
	*	Multiplication	x = x * y;
	/	Division	x = x/y;
	%	Modulus	x = x % y;
Logical operators	&&	Logical AND	x = true && false;
	‖	Logical OR	x = true ‖ false;
	&	Bitwise AND	x = x & 0xFF;
	‖	Bitwise OR	x = x ‖ 0xFF;
	~	Bitwise NOT	x =~ x;
	!	Logical NOT	false = !true
	≫	Shift bits right	x = x ≫ 1;
	≪	Shift bits left	x = x ≪ 2;
Equality operators	==	Equal to	if(x == 10) { . . . }
	!=	Not equal to	if(x != 10) { . . . }
	<	Less than	if(x < 10) { . . . }
	>	Greater than	if(x > 10) { . . . }
	<=	Less than or equal to	if(x <= 10) { . . . }
	>=	Greater than or equal to	if(x >= 10) { . . . }

1.4.6 Conditional Statements and Iteration

To control the flow of execution in a C program we usually use a conditional 'if' statement as well as 'for' and 'while' loops.

The 'if' statement is used to allow decisions to be made about parts of the program that will be executed depending on some conditions tested by the 'if' statement in the program. If the condition given to the 'if' statement is true then a section of code is executed as outlined below:

```
if(condition)
{
   execute this code if condition is true
}
```

An example of a simple usage of the 'if' statement is given below:

```
if(x<0)
{
   x = -1 * x;
}
```

Sometimes you might want to perform one action when the condition is true and another action when the condition is false. Here, an 'else' statement needs to be combined with the 'if' statement:

```
if(condition)
{
   execute this code if condition is true
}
else
{
   execute this code if condition is false
}
```

It is easy to chain a lot of 'if–else' statements:

```
if(condition 1)
{
   execute this code if condition 1 is true
}
else if(condition 2)
{
   execute this code if condition 2 is true and condition 1
   is false
}
```

```
else // optional
{
   execute this code if condition 1 and condition 2 are false
}
```

It is also possible to nest 'if' statements:

```
if(condition 1)
   if(condition 2)
   {
      execute this code if condition 1 and condition 2 are true
   }
   else
   {
      execute this code if condition 1 or condition 2 are false
   }
```

While the 'if' statement allows branching in the program flow, 'for', 'while' and 'do' statements allow the repeated execution of code in 'loops'. It has been proven that the only loop needed for programming is the 'while' loop but sometimes it is more convenient to use the 'for' loop instead. While the condition specified in the 'while' statement is true a statement or group of statements (compound statements) are executed as outlined below:

```
while(condition)
{
   execute this code while condition is true
}
```

An example of the simple usage of a 'while' statement is given below:

```
while(x>0)
{
   result = result * x;
   x = x - 1;
}
```

The other type of loop is the 'for' loop. It takes three parameters: the starting number, the test condition that determines when the loop stops and the increment expression.

```
for(initial; condition; adjust)
{
   code to be executed while the condition is true
}
```

An example of the simple usage of a 'for' statement is given below:

```
for(counter = 1; counter <= x; counter = counter +1)
{
   result = result * x;
}
```

1.4.7 Example C Program

To start programming the PIC using C language, we will develop a program very similar to the one designed using the assembler language shown in Listing 1.1. The program will still generate two different bit patterns on port B but will be slightly more complex. In order to 'slow down' the program and keep each of two bit patterns on port B for a longer period of time, an additional delay will be generated in the program. This program is given in Listing 1.5.

```
1 // this program outputs two binary patterns on port B
2 // one after the other - 01010101 and 10101010 — with
3 // some delay between them
4
5 #include <pic16f62xa.h>
6
7 void delay(void);
8
9 void main(void)
10 {
11      int j;
12
13      TRISB = 0x00;       // make port B, all bits, outputs
14      while (1)           // do forever, i.e. always true
15      {
16          PORTB = 0x55;   // port B = 01010101
17          for(j=0;j<10000;j++)     // produce a delay
18          {
19          }
20          PORTB = 0xAA;   // port B = 10101010
21          delay();
22      }
23 }
24
25 void delay(void)
26 {
27      int j;
28      for(j=0;j<10000;j++)   // produce a delay
29      {
30      }
31 }
```

Listing 1.5   First C PIC program

The 'main()' function of this program is defined between lines 9 and 23. After configuring all pins of port B as output pins on line 13, the program enters the 'while(1)' loop to execute the sequence of statements in this loop. This type of loop is called an infinite loop. The condition of this loop is always true (it is 1) so the program will stay in this loop 'forever', executing statements in the loop as long as there is a power supply to PIC. Unlike conventional C programs, embedded microcontrollers usually run their software as an infinite (do forever) loop. As long as the larger system incorporating the microcontroller operates, it will need the microcontroller to perform its specific task. This task is described by the C sequence enclosed in '{}' following the while(1) statement.

Another way to implement the infinite loop is to use the 'for' statement without specifying any condition, i.e. 'for(;;)'.

Having entered the infinite loop, the program first sends the hexadecimal value 55h to port B. After that it enters the 'for' loop at line 17. This loop does nothing particularly useful as the curly brackets following the 'for' statement contain no other statements. All it does is waste some time incrementing the loop counter j before the next hexadecimal value (AAh) is output to Port B on line 20. Loop counter j is declared as integer at the beginning of the main program on line 11.

Line 21 calls the function 'delay()', which is declared in line 7 and defined between lines 25 and 31. This function, like the 'for' loop in the main program, does nothing else but wastes some more time before the program branches back to line 16 to output value 55h to port B again. In fact the main body of this function contains just another 'for' loop, identical to the 'for' loop already executed in the main program. This is certainly not very good programming practice. A more experienced programmer would probably call the delay() function after line 16 in the program to follow the PORTB = 0x55 statement and generate the required time delay. The 'for' loop used in the main program will then become obsolete. This would reduce the size of the program and make the program somewhat easier to read and understand. We decided to use this rather clumsy approach to demonstrate the use of functions in the C program and to compare the performance of the delay() function with the delay generated by the 'for' loop in the main program. With 4 MHz crystal, both time-wasting approaches – the 'for' loop in the main program and the delay() function – can be expected to generate a delay of approximately 150 ms. A method to measure the exact delays generated in this program will be explained towards the end of this chapter.

1.5 MPLAB Integrated Development Environment (IDE)

MPLAB is a Windows-based program that makes writing, developing and debugging programs for PIC microcontrollers an easier task. This development environment is produced by Microchip but can incorporate third-party software tools that can also be used when developing an embedded PIC-based application. MPLAB is relatively easy to use, has a friendly graphical interface and can be freely downloaded from the Microchip web site. This free version can later be easily upgraded with more powerful C compilers and other additions for more complex tasks.

Installation of this package is a relatively straightforward procedure. The MPLAB installation program is provided on the web site for this book but can also be downloaded directly from the Microchip web site (http://www.microchip.com/, accessed 29 December 2008). The reader should bear in mind that this software is updated frequently by the manufacturer so the most recent version of the MPLAB software will always be available from the Microchip web site and might have a slightly different set of features from those described in this book. MPLAB is, however, backwards compatible, which means that all programs and features described in this book should also be available and accessible in any newer version of MPLAB. In case the reader is unsure and does not want to experiment with newer versions of MPLAB, the best approach is to stick with version 8 of MPLAB, provided with this book, download it from the companion web site and install it on the computer.

The installation procedure is relatively simple. A zip file, 'MPLAB_v8.zip', containing all necessary files for MPLAB installation needs to be extracted, preferably in an empty or newly created directory. Double clicking on the extracted installation file 'Install_MPLAB_v8.exe' will start Installation Shield and bring up the MPLAB installation window.

Nothing special needs to be configured for the MPLAB installation. All that needs to be done is to accept the default settings by choosing the option NEXT several times and accepting the terms of agreement. After several minutes, MPLAB should be installed on your computer without any problems. At the end of the MPLAB installation, you will be offered an option to add the HI-TECH PICC Lite software suite to your MPLAB. This is a free version of C compiler produced by the third-party supplier HI-TECH. Most of the C programs from this book are compiled using this software, so it is advisable to accept this option and install PICC at this point. If you do not do it at this point, the HI-TECH

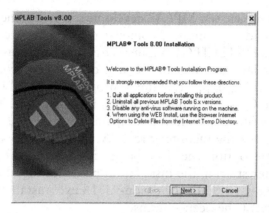

Figure 1.5 Start of the MPLAB installation procedure

© Microchip Technology Inc. Reproduced with permission.

PICC compiler can easily be downloaded from the book web site and installed at some later time

Select NEXT and accept the terms of agreement and select NEXT again. Accept the options offered for the installation and click NEXT again to complete the installation procedure. At this point you will have the option to complete installation and restart the computer. Accept this option if you want to use MPLAB immediately. If not, you will need to restart your computer at some later time before starting to work with

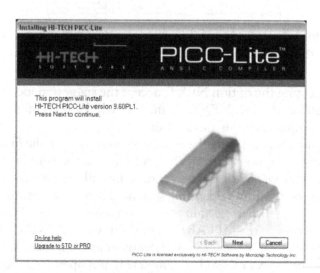

Figure 1.6 Addition of the HI-TECH PICC compiler to the MPLAB suite

© Microchip Technology Inc. Reproduced with permission.

MPLAB. To complete our MPLAB development environment it might be a good idea to add one more C compiler to it – C18 C compiler.

The installed HI-TECH C compiler is free and can be used to compile programs for a large number of PIC microcontrollers from the PIC10, PIC12 and PIC16 group. To compile programs for PIC18 devices we need to use the C18 C compiler provided by Microchip. This compiler can also be downloaded from the companion web site for this book or, alternatively, from the manufacturer's (Microchip) web site. It installs easily but some options need to be specified during the installation procedure. After starting the installation procedure by double clicking on the 'MPLAB-C18-Student Edition-v3_16.exe' installation file, click NEXT and accept the licence terms.

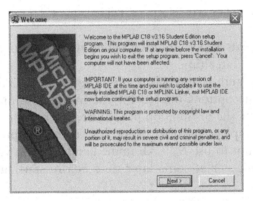

Figure 1.7 Start of the C18 compiler installation procedure
© Microchip Technology Inc. Reproduced with permission.

After selecting the option NEXT, accept the suggested location of the compiler files and select NEXT again. You will be presented with three screens of options for this installation.

We recommend accepting the suggested options from the first window and selecting all options offered (but not selected for you) in the second and third windows. With this selection you will set the environmental variables in the Windows operating system, which makes it easier for the compiler to locate where certain files are on the computer. You will also add this compiler to the MPLAB environment. Clicking NEXT again will start the installation of the C18 compiler on your computer. At the end of the installation, select FINISH to complete it and open the Release Notes for C18, if you want to read through them at this point. Do not forget to restart your computer before starting to use MPLAB for the first time.

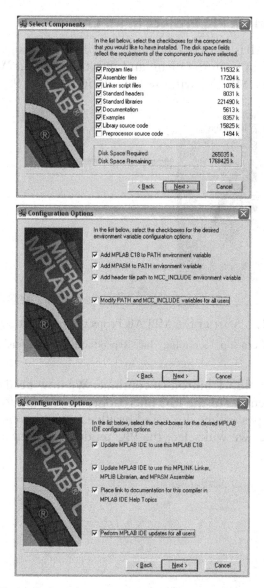

Figure 1.8 Selecting C18 installation options

© Microchip Technology Inc. Reproduced with permission.

After successful installation and restarting the computer, the PIC program can be developed using MPLAB in three main steps – MPLAB project specification, program writing and conversion of the written program into executable binary (1-0 code understandable by the microcontroller).

1.5.1   Creating a PIC Project in MPLAB

Once the MPLAB is started, to create a new MPLAB project click on the PROJECT option from the menu and select the Project Wizard, which will open a new MPLAB window, shown in Figure 1.9.

Figure 1.9 Start of the MPLAB Project Wizard procedure

© Microchip Technology Inc. Reproduced with permission.

Click on the NEXT button to continue. We now need to choose the appropriate member of the PIC microcontroller family for this project. Various PIC types will be discussed briefly later in this chapter. For our first PIC project we will select the PIC model discussed in this section: PIC16F627A.

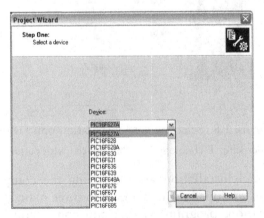

Figure 1.10 Selecting the PIC model during the project definition

© Microchip Technology Inc. Reproduced with permission.

In the next step of the project specification procedure we need to define the programming language to be used in the project. We will use assembler language for our first project, so we need to select the MPASM toolsuite for assembler language software as shown in Figure 1.11.

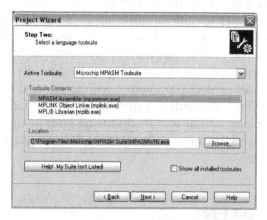

Figure 1.11 Selecting a language and suitable tools for project

© Microchip Technology Inc. Reproduced with permission.

In the final step, we need to select the name and folder for our project. Usually, we select the project name to somehow reflect the nature and purpose of the program. It is usually a good idea to create a new folder for each independent MPLAB project. This is going to be our first PIC program, so we have created a folder 'E:\PICprograms\firstprogram' to hold our new PIC project called 'firstPICprogram'.

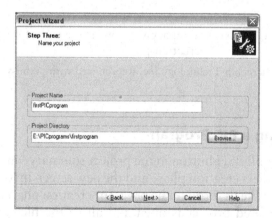

Figure 1.12 Specifying project name and location

© Microchip Technology Inc. Reproduced with permission.

We can now click on the NEXT button to finalize the project specification procedure. A new window will offer the possibility of including some existing files in the newly created PIC project. We are going to create a new assembler file for this project so we click NEXT at this stage. This will bring on a new project summary window that contains the summary specification of our project. At any stage of the project specification procedure it is possible to step back by clicking on the BACK button in order to change parameters set in that stage. The same option exists in the project summary window. By clicking on the FINISH button, all project parameters will be selected and the project finally created. This will result in a number of project files being created in the project folder, which will be discussed later in the book, but we will now proceed to the second phase of the PIC project development – program writing.

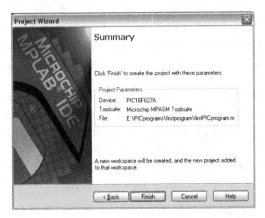

Figure 1.13 Project summary window at the end of the project creation procedure

© Microchip Technology Inc. Reproduced with permission.

1.5.2 Writing a PIC Program

By clicking the FINISH button in the project summary window we have finished the project creation phase and the new screen in MPLAB should now appear to take us through the program-creation phase. At this point your newly created project is open and all of the files created in the previous phase and associated with this project are automatically added to a project.

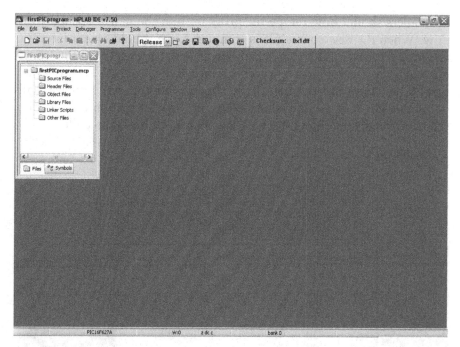

Figure 1.14 Main project window at the start of source file creation

© Microchip Technology Inc. Reproduced with permission.

To create a new file to hold our assembly language program we need to select FILE and then NEW from the MPLAB menu. This will open a new window. In this window we will type assembly code for our PIC program. This code is usually called source code and the file holding the source code is called the source file. Every PIC project needs to contain at least one source file.

We can now start typing the source code into the source file window. The program that we will enter at this stage is a simple program developed in Section 1.3.3 and will only serve the purpose of demonstrating the use of the MPLAB development environment. This program can be entered into this window manually, which is a recommended option for the reader at this stage. Alternatively, all of the programs from this book are also available from this book's companion web site. They can be downloaded, copied and pasted into the source window or once downloaded into an appropriate directory on your, computer, included straight into the project. This will be a recommended option for more complex programs used in the later chapters of this book.

Readers should now retype the program into the newly opened window. When the new source file is completed, we should save it into our working folder (E:\PICprograms\firstprogram), for example, using some meaningful name for it, to reflect the nature of the program. As this is our first PIC program and the name of the project is 'firstPICprogram' we shall name this file 'firstprogram.asm'. Note the 'asm' extension for this file, which indicates the assembler language nature of this source file.

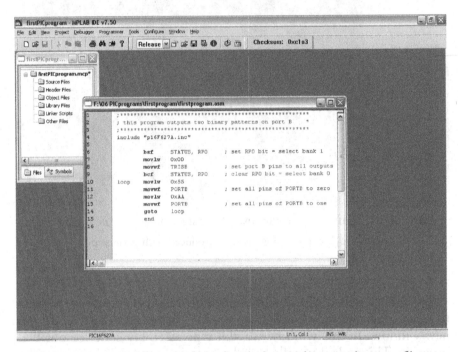

Figure 1.15 Main project window with the newly created source file

© Microchip Technology Inc. Reproduced with permission.

The new file, 'firstprogram.asm', should be added to our project. To do this, we need to select the ADD TO THE PROJECT option from the PROJECT menu. A new browser-type window will open. We now need to find our PIC folder and select the file 'firstprogram.asm', as shown in Figure 1.16.

The addition of the new source file 'firstprogram.asm' to our project will be reflected in the MPLAB project window shown in Figure 1.17. It can be seen that 'firstprogram.asm' file now appears in the source files group in that window.

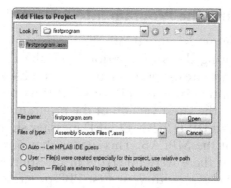

Figure 1.16 Adding source file to a current project

© Microchip Technology Inc. Reproduced with permission.

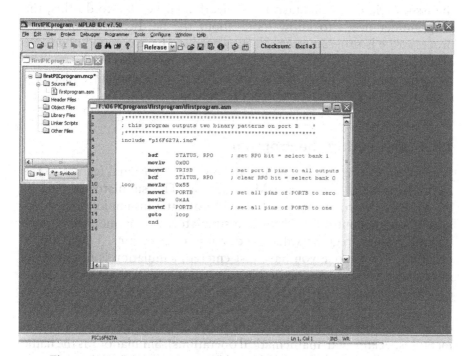

Figure 1.17 Main project window with the newly created source
file added to the project

© Microchip Technology Inc. Reproduced with permission.

1.5.3 Translating the Program into Executable Code

Once the full program is entered into the source code window we can resave it by choosing the SAVE option from the FILE menu. Alternatively we can translate the program into executable form by selecting the BUILD ALL option from the PROJECT menu or just pressing F10. This will also save the latest version of the program. If the program is written properly, a BUILD SUCCEEDED message should appear in the output window of the MPLAB. This means that our translation was successful and that there were no errors in the source code we typed. This should be the case with your first PIC program if you typed it correctly.

If a syntax error does show up, it needs to be corrected. By double clicking on the relevant error message in the output window you will be transferred to the source code window in the assembler code line where the error was detected. Understanding what the particular error is and how to correct it is one of the main tasks facing the programmer and the rest of this book will hopefully help the reader in understanding some of the issues related to this task more clearly. To be able to do this, the programmer needs to know more about the hardware of the target processor he is programming. He also needs to have a good knowledge of the programming language he is using to program this hardware. The program you have just entered should be simple enough and errors should be easily detected using MPLAB help.

1.5.4 Simulating the Program

We will finish this section by introducing one additional feature of the MPLAB package – the MPSIM simulator. The simulator is part of the MPLAB environment, which provides a better insight into the workings of a microcontroller. Through a simulator, we can monitor current variable values, register values and status of port pins. For a simple program, like the one you have just entered, simulation is not of great importance as the operation of this program is quite straightforward and easy to understand even for the PIC beginner. However, the simulator can be of great help with more complicated programs which include timers, different conditions where something happens and other similar requirements (especially with mathematical operations). Simulation, as the name indicates, 'simulates the work of a microcontroller'. As the microcontroller executes instructions one by one, the simulator moves through a program step-by-step (line-by-line) and follows what goes on with data within the microcontroller. When writing is completed, it is advisable to first test the program in a simulator and then run it in a real situation. Unfortunately, as with many other good habits, many programmers tend to avoid this one too, more or less. Reasons for this are partly personality and partly a lack of good simulators. We will now explain how to use the MPSIM simulator to simulate and test our first PIC program.

Once our assembler program is successfully built (assembled) we need to select the MPSIM debugging tool from the SELECT TOOL option from DEBUGGER menu. A new set of icons will appear on the menu bar of MPLAB and additional menu items will appear in the debugger menu.

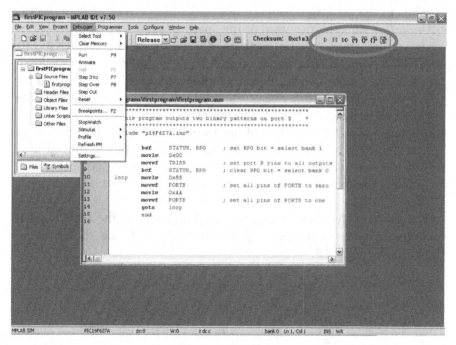

Figure 1.18   Main project window with the source file and invoked simulator

© Microchip Technology Inc. Reproduced with permission.

Our program is now ready to run. It is usually a good idea to reset the program before a fresh run. This can be done by selecting the RESET option from the DEBUGGER menu. A green arrow should appear at the left margin of the source-code window indicating that the first line of the code is to be executed. By selecting RUN from the DEBUGGER menu the program will start running. A text message 'Running . . .' will appear on the status bar. To halt the program execution we need to select HALT from the DEBUGGER menu. The line of code where the application halted will be indicated by the green arrow.

We can also execute a program through single steps – executing each instruction from the program individually. To single-step through the application program, select the STEP INTO option. This will execute the currently indicated line of code and move the arrow to the next line of code to be executed. Options for program execution using MPSIM are:

- *Run* – runs the program at the simulation's full speed but variables cannot be watched in this mode; this option is normally used with breakpoints set in the program to stop the execution at one or more points in the program.
- *Animate* – runs the program at the speed of only several instructions per second (rate can be further adjusted from the DEBUGGER-SETTINGS menu); the program execution is therefore slowed down so that the effect of each line can be observed.
- *Step into* – executes one line of the program and steps into subroutine if the call to subroutine is encountered.
- *Step over* – executes one line of the program and subroutine in one go if the call to subroutine is encountered (it steps over the subroutine).
- *Step out* – when the program is in a subroutine this option will complete the execution of the subroutine at full speed.

The shortcuts for these commonly used functions in the DEBUGGER toolbar can be used to speed up this procedure. Those shortcuts are shown in Figure 1.19.

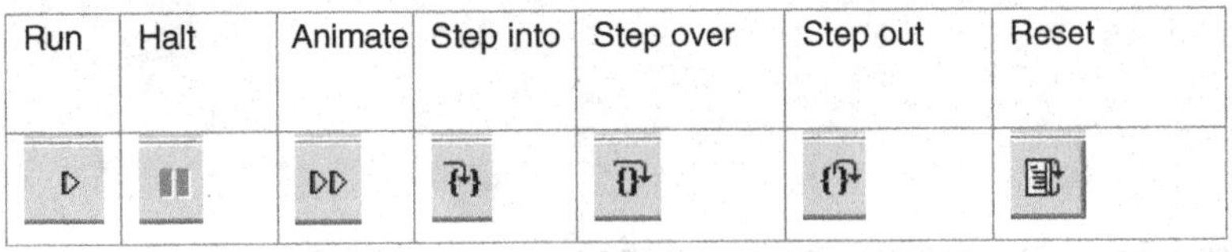

Run	Halt	Animate	Step into	Step over	Step out	Reset
▷	❚❚	▷▷	⁐	⁐	⁐	⁐

Figure 1.19   MPSIM shortcuts

© Microchip Technology Inc. Reproduced with permission.

Another feature of MPSIM is the ability to track changes in variable or register values in the program and observe those changes during the program execution. This can be done using the Watch window. A number of Watch windows can be activated from the VIEW menu. In our current application we might want to observe more closely the value of PORTB. This is one of the PIC special function registers, so we can

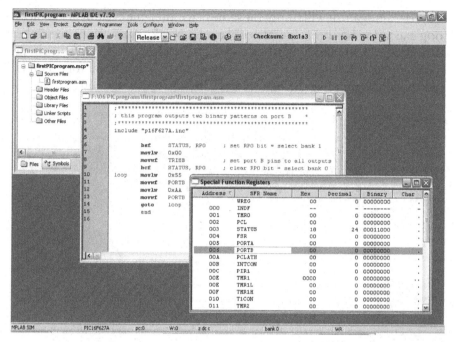

Figure 1.20 Main project window with the SFR Watch window

© Microchip Technology Inc. Reproduced with permission.

select the SPECIAL FUNCTION REGISTERS option from the VIEW menu. A new window with all the files appears in MPLAB.

We can now run the program by animating it or by single-stepping through it. Change on PORTB can be observed clearly in both cases. If we decide just to run the program, those changes cannot be tracked as this option runs the program in 'near real time', which is too fast for us to spot anything if the real system is used and too fast for MPLAB to update the Watch window. This feature can, however, be used in combination with breakpoints in the program to reach a certain point in the program quickly, stop the execution of the program and observe the state of the registers or memory at that point in the program. To set the breakpoint in the program you need to double-click on that line of the program in the source window. Let us set the breakpoint on the last line of our program (line 14) by double clicking on that line. A 'stop'-type breakpoint symbol should appear next to the selected line of code in the left margin of the source window. Now, reset the program and choose the option 'run' from the DEBUGGER toolbar. The program will execute quickly and stop at line 14 of our program as indicated by the green arrow over the breakpoint sign in the source window. If we run the program again, it will stop at the same place after one cycle of our endless loop execution has been completed. Several breakpoints can be set in the same program. Breakpoints are a good way to get to a certain point in a long and complicated program quickly. They are not so important for a simple and short program like the one we have in our source window.

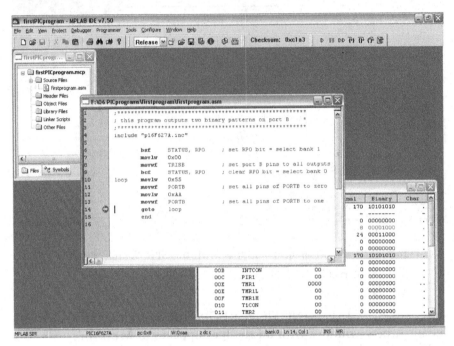

Figure 1.21 Program stop at the breakpoint

© Microchip Technology Inc. Reproduced with permission.

Another useful option in MPSIM is the possibility of measuring the execution time of our program or parts of the program. To do this we first need to set the frequency of the simulator clock. This can be done by selecting the SETTINGS option from the bottom of the DEBUGGER menu. Set the value of the processor frequency to 4 MHz and close this window.

Now select the option STOPWATCH from the same menu to bring up the stopwatch window. Reset the simulation and run the program again. Measurement in the stopwatch window will show the time of the execution of one loop sequence in our program as the program will again stop at the breakpoint we set in the previous phase of program testing. The elapsed time is $8 \, \mu s$. This is the correct measurement and agrees with our estimation for the program time execution on PIC made in Section 1.2.4. Our loop consists of eight instructions and each instruction took $1 \, \mu s$ to

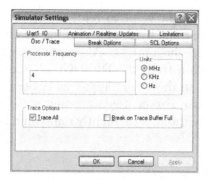

Figure 1.22 Selecting the frequency of the processor for the simulator

© Microchip Technology Inc. Reproduced with permission.

execute. To confirm this, we can reset the program and start stepping through it while observing the stopwatch window. Each step through the program will increment the stopwatch window by 1 μs. Notice that the only deviation from this rule is the execution of the goto instruction in our program. This is a flow-control type instruction and, like the other flow-control instruction (call), it takes two clock cycles to execute, which is 2 μs execution time for the 4 MHz clock.

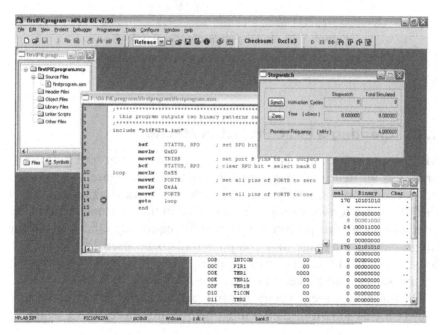

Figure 1.23 Stopwatch window with the indicated time of program execution (up to a breakpoint in the program)

© Microchip Technology Inc. Reproduced with permission.

At this point you might want to remove the breakpoint from your program. This can be done by double clicking again on the breakpoint line or break sign in the margin. Alternatively you can choose the BREAKPOINTS option from the DEBUGGER menu and remove the breakpoint by filling in the newly appeared dialogue box. This break-points dialogue box can also be used to configure (set new, remove some or all or just disable without permanently removing) breakpoints in your program.

Figure 1.24 Breakpoints control window

© Microchip Technology Inc. Reproduced with permission.

1.5.5 Creating a PIC C Project in MPLAB

MPLAB can be used in a similar way to create and test project for the PIC programs written in C programming language. To create a C-type PIC project after opening MPLAB, start a Project Wizard again and select the PIC device for your project as described in the previous section. The next option in the project creation will require specification of the programming language and compiler tool to be used in the project. Select the 'HI-TECH Universal Toolsuite' and 'HI-TECH C Compiler' on this screen and proceed by selecting NEXT.

On the next screen specify the full path and the name of the new project file as shown in the window below and select NEXT again.

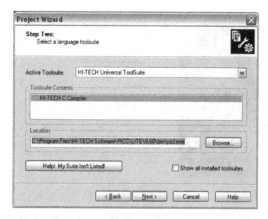

Figure 1.25 Project Wizard window with selected C compiler and specified project location

© Microchip Technology Inc. Reproduced with permission.

Figure 1.26 Project Wizard ready for the creation of the new project

© Microchip Technology Inc. Reproduced with permission.

On the next screen select NEXT again and FINISH on the project summary window.

Program creation is very similar to the process for the assembler language program explained in Section 1.5.2. This time we need to type the C version of our first PIC program given in Listing 1.5 and save it in the file with a .c extension. The C source file needs to be added to our project and the project needs to be built and tested in a similar way to the assembler MPLAB project. Figure 1.28 shows MPLAB with a project

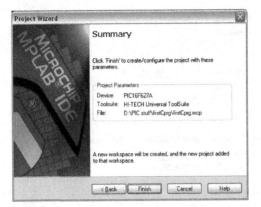

Figure 1.27 Project summary window at the end of the C project creation
© Microchip Technology Inc. Reproduced with permission.

created and built using the C program from Section 1.4.7. The MPLAB
SIM tool is activated and PORTB is selected in the Watch window. Two
breakpoints are set in the program to stop the execution of the program
so that the change of bit pattern on PORTB can be observed.

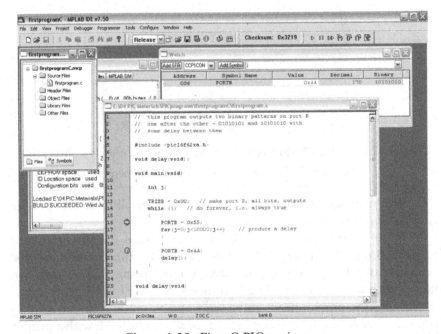

Figure 1.28 First C PIC project
© Microchip Technology Inc. Reproduced with permission.

Time delays generated by the 'for' loop and the delay() function can be measured using those breakpoints and a Stopwatch feature from the DEBUGGER menu. To do this, first set the Processor Frequency to 4 MHz by selecting the Settings . . . option from the DEBUGGER menu. In the new Simulator Settings window select Osc/Trace tab and set the Processor Frequency to 4 MHz. The corresponding screenshot of MPLAB is shown in Figure 1.29.

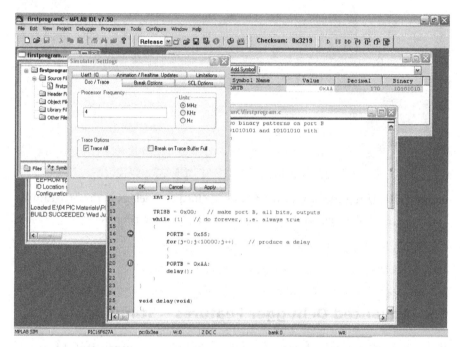

Figure 1.29 Setting the processor frequency

© Microchip Technology Inc. Reproduced with permission.

Confirm your selection by clicking OK to close the Simulator Settings dialogue box and select the Stopwatch option from Debugger menu. Run the program to the first breakpoint and reset the stopwatch time by pressing Zero. Now, run the program again to the second breakpoint and note the elapsed time – 140 ms. Zero and run again. This time the program was delayed by the delay() function and measured delay was slightly longer – 150 ms. Can you think of any reasons for this? The number of instruction cycles executed to the first and second break-points, provided by the Stopwatch window, can give you a clue.

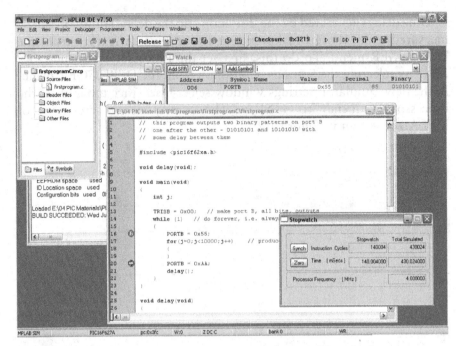

Figure 1.30   Using the Stopwatch feature to measure delays in the program

© Microchip Technology Inc. Reproduced with permission.

1.6 Advanced Debugger Features – Stimulus

We will complete this chapter by demonstrating one more useful feature of the MPLAB SIM tool – simulation of program stimulus from the outside world. To do this we will modify our C program according to Listing 1.6.

The modified program first reads the state of PORTA on line 13 and stores the result in a variable called 'temp' – unsigned character. According to the value of this variable (the state of PORTA), one of two binary patterns will be output to PORTB – 55h if all pins of PORTA are set to logic low (0) or AAh if any of PORTA pins is set to logic high (1). The test is implemented using if–else statement between lines 15 and 22. Pins of PORTA are configured as inputs at the beginning of the program, on line 9.

```
 1 // this program outputs one of two binary patterns on port B
 2 // depending on the state of port A
 3
 4 #include <pic16f62xa.h>
 5
 6 void main(void)
 7 {
 8      unsigned char temp;
 9      TRISA = 0xFF;           // make port A, all inputs
10      TRISB = 0x00;           // make port B, all bits, outputs
11      while (1)               // do forever, i.e. always true
12      {
13           temp = PORTA;
14
15      if (temp == 0)
16      {
17           PORTB = 0x55;
18      }
19      else
20      {
21           PORTB = 0xAA;
22      }
23   }
24 }
```

Listing 1.6 Second C PIC program

Create a new PIC project, type in the C program given in Listing 1.6, add it to the project and build the project. Open the Watch window and select PORTB and PORTA to watch during the program animation. In order to test the operation of this project we would need to somehow change the state of the PORTA pins during the program animation. This can be done using the stimulus feature of the MPLAB SIM tool. To open the Stimulus window, select Debugger-Stimulus-New Workbook from the MPLAB menu. The Stimulus window should appear in MPLAB as shown in Figure 1.31.

Three main types of stimulus can be set up in MPLAB using the Stimulus window:

- *Manual triggers* – changes in digital signal levels caused by clicking on a button with a mouse during the program animation or while single-stepping through the program. These allow you to simulate the action of closing a switch, or pulsing a pin.
- *Sequential data* – applied to pins, registers, or bits in registers from a predefined list.

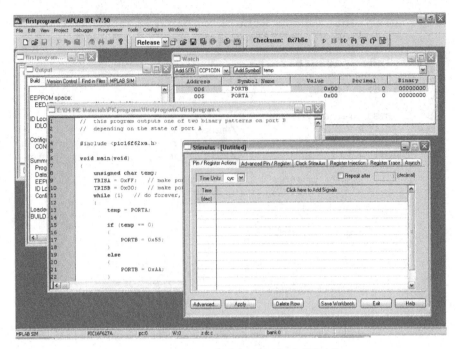

Figure 1.31 Starting Stimulus in MPLAB

© Microchip Technology Inc. Reproduced with permission.

- *Cyclic stimulus* – repeating waveform for a predetermined length of time or continuously.

In this chapter we will show how to set up and use the first two types of stimulus, leaving the cyclic type stimulus for later chapters. We will only demonstrate the basic configuration for those two stimulus types. It is left to the reader to investigate this feature of the MPLAB SIM tool further.

A manually triggered stimulus is an asynchronous type of stimulus, so to configure it we need to select the Asynch tab from the Stimulus window. By clicking on the first line of the Pin/SFR column in the table under this tab we can select the pin or register to be affected by the stimulus. In our example we use PORTA as an input port, select RA0 – first pin of PORTA, to be manually set during the animation of the program. By clicking under the Action field we can select one of five possible types of stimulus for RA0 – 'Set High', 'Set Low', 'Toggle', 'Pulse High' or 'Pulse Low'. Select – 'Set High'. By pressing the first button in the same row (under the Fire label in the same table) during the program animation we will be able to set RA0 high. Without further action RA0 would remain high until the end of the animation. It might

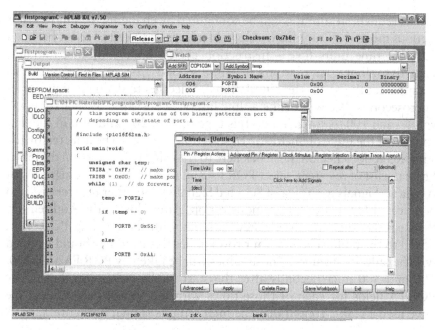

Figure 1.32   Configuration of asynchronous stimulus in MPLAB

© Microchip Technology Inc. Reproduced with permission.

be a good idea to add a second action in order to set this pin low during the same testing sequence. To do this, go to the second row of this table, select RA0 again but choose 'Set Low' as the action in this case. This setup is shown in Figure 1.32. Fire buttons used to set or reset pin RA0 during the program animation are highlighted in this figure.

All that needs to be done now is to animate the program by pressing the ⏸ button on the MPLAB shortcut bar or by selecting the Animate option from the DEBUGGER menu. The program will run in the slow mode so its action can be clearly observed. Once the program enters the infinite loop it circulates between lines 11 and 17 – since all of the PORTA pins are initially 0, the condition following the 'if' statement is true and the program continuously sends 55h to PORTB. This can be checked in the Watch window where PORTB = 0x55 while PORTA = 0x00. Firing the first event – setting RA0 high – will change this sequence. PORTA will change its state to 0x01 and the program will send AAh to PORTB instead of 55h. This can again be checked in the Watch window. Firing the second event during the same animation cycle will set RA0 low. The result in the Watch window is: PORTB = 0x55 and PORTA = 0x00 again. More pins of PORTA and different actions can be added to this basic configuration to make it more interesting and versatile.

The sequential stimulus can be configured by changing the Pin/Register Actions tab in the Stimulus window. Here, we first need to select between different time units (cyc – instruction cycles, h:m:s – hours, minutes, seconds for longer simulations, ms – milliseconds, μs – microseconds or ns – nanoseconds for shorter simulations). Select μs to define a sequence in microseconds. We will define a time interval consisting of four subintervals – each 100 μs long. To do this type 0, 100, 200 and 300 in the first column of the table (Time/dec). To select the register to be affected by the stimulus action click on the 'Click here to Add Signals' heading in the table. From the list of registers, select PORTA and Add to move it from the list of Available Signals to the list of Selected Signal(s). No other register needs to be selected for this simple demonstration. After closing the Add/remove Pin/Registers dialogue box a new column with the heading PORTA is added to the table. The first four rows under this heading should be filled with hexadecimal values. Those values will be taken by PORTA at 0, 100, 200 and 300 microseconds after the start of the animation. Fill in the PORTA fields for each of those times with the appropriate values. We have used FF, 00, 10 and 00. The final setup is shown in Figure 1.33. A

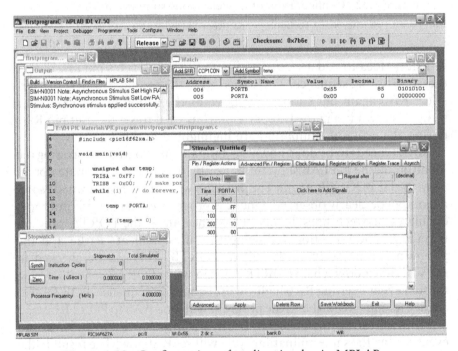

Figure 1.33 Configuration of cyclic stimulus in MPLAB

© Microchip Technology Inc. Reproduced with permission.

Stopwatch feature is also activated to help in following the execution time of the program. Before starting the animation of the program apply this sequence by clicking on the Apply button in the Stimulus window.

Start the animation of the program and observe the changes in the Watch window at specified times. PORTA is changing the value according to specification in the Stimulus window and PORTB is following these changes by switching between two patterns after $100\,\mu s$, $200\,\mu s$ and $300\,\mu s$. The only slightly confusing behaviour might be the value of PORTA during the first $0{-}100\,\mu s$ time interval. Although we set it to FFh, the Watch window shows PORTA $= 0 \times 3F$! If you have not noticed it, try resetting the program and animating it again. Can you explain this anomaly? If you can, it is time to start with some simple PIC projects in Chapter 2.

CHAPTER

2

Simple Interfaces

2.1 Introduction

Two circuits, one based on the PIC12F629 (Figure 2.1) and the other based on the PIC16F627A (Figure 2.2), will be used in this chapter to develop applications involving simple interfaces between those PIC microcontrollers and the outside world.

The PIC12F629 is an eight-pin dual-in-line (DIL) microcontroller. Six of the eight pins of this microcontroller can be used as input/output (I/O) pins. As shown in Figure 2.1 these pins have other functions but they are used as I/Os for all the examples given in this chapter. This chip only requires a power supply to be connected to pin 1 (V_{DD}) and pin 8 (V_{SS}) to function as a complete microcontroller and there is no need for an external crystal clock or reset circuitry. If external crystal or reset circuitry is required by the application, then it is possible to include this by using pins 2 and 3 for the crystal clock and pin 4 for the reset circuit, but doing so would reduce the number of I/O pins available for the application. The programming mode of the PIC chip determines how the microcontroller is to be configured.

The PIC16F627A is an 18-pin package with 16 available I/O lines. This device also offers a four-channel analog-to-digital converter (ADC)

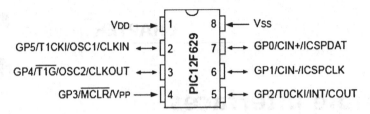

Figure 2.1 The PIC12F629 microcontroller

© Microchip Technology Inc. Reproduced with permission.

on pins 17, 18, 1 and 2. Again the 16 I/O lines can be configured to perform other functions, such as detailed in Figure 2.2.

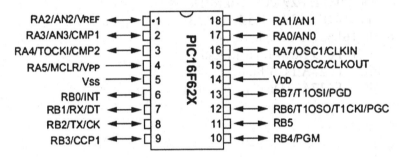

Figure 2.2 The PIC16F627A microcontroller

© Microchip Technology Inc. Reproduced with permission.

We shall first design an appropriate circuit using the PIC12F629, show the construction of the circuit using a strip board and then develop a PCB board to implement the same circuit. Programs to make use of this board in C and assembler language will then be developed and their operation explained in detail. We will repeat the same procedure (with slightly more complex programs) for the circuits and programs using the PIC16F627A microcontroller.

2.2 PIC12F629 Circuit Design

Programs developed for the PIC12F629 microcontroller will use one or two digital outputs to switch light-emitting diodes (LEDs) on and off and one digital input to read the state of the push button. For build simplicity, we chose GP4(3) as an input (to read a push button state) and GP1(6) and GP2(5) as outputs (to drive two LEDs connected to those pins). The complete circuit is shown in Figure 2.3.

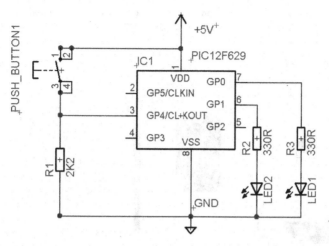

Figure 2.3 The PIC 12F629 microcontroller schematic

The input pin GP4 is normally connected to the ground (0 V) through R1 and is read as logic 0 by the program inside the PIC. When the push button is pressed, the GP4 pin is connected directly to the +5 V supply voltage and the input is read as a logic 1 by the program. The PIC I/O pins can supply a current of 25 mA (sourcing current) to a load or sink 25 mA (sink current) from a load. In this case we are sourcing the currents to the LEDs and we need to calculate the value of the limiting resistors R2 and R3. The LED requires 10 mA through it and about 1.5 V across it, to be bright enough. To switch on the LED we need to set the corresponding output pin of the PIC microcontroller to a logic 1. This provides about 5 V on the corresponding pin, so the value of resistors R2 and R3 can be calculated as follows:

$$V_{supply} = R2 \times I_{LED} + V_{LED}$$
$$5\,V = R2 \times 10\,mA + 1.5\,V \Rightarrow R2 = 350\,\Omega \qquad (2.1)$$

We therefore choose an available resistor value of 330 Ω for resistors R2 and R3.

2.3 The PIC12F629 Strip Board Design

Here, for ease of constructing a board, we have used a strip board with 13 rows and 25 columns (see Figure 2.4). Rows B and L are assigned to +5 V track and to 0 V track respectively. The PIC 12F629 is placed on rows F to I and columns 9 to 12. The tracks F to I on column 10 need to

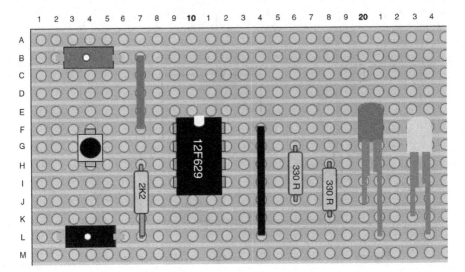

Figure 2.4 PIC12F629 microcontroller strip board design

be cut to make sure there is no connection between the left and right side of the chip (pin 1 must NOT be connected to pin 8). Note that there is a link from the B to the F track on the left side of the PIC to connect the +5 V supply to the +5 V of the PIC and also a link from the L to the F track on the right side of the PIC to connect 0 V (ground) to 0 V of the PIC. A push-button switch is connected between the F and H tracks to connect +5 V to pin 3 (GP4) of the PIC. An R1 (2K2) resistor is placed between pin 3 on the H track and Ground on the L track. Resistors R3 and R2 are placed between tracks G and J, H and K respectively. The anode (+ ve) of LED 1 is connected to track J and its cathode (−ve) to the L track (Ground). The anode of LED 2 is connected to the K track and its cathode to the L track.

2.4   The PIC12F629 PCB Board Design

To draw the schematic, a CAD package 'EAGLE 4.16 r2 Lite' was used. This is a free (lite) version of the software with some limitations on the size of the board and the number of layers but it meets the requirements for all the PCB designs in this book. The schematic was put through all the processes necessary to obtain the PCB design shown in Figure 2.5. Here we used only a single-layer board with the components on top and the track side underneath. This board is produced not to teach PCB

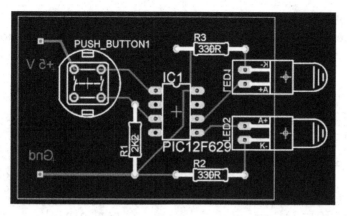

Figure 2.5 PIC12F629 microcontroller PCB board design

design but to provide a quick and simple board. Note here that the +5 V
and Gnd labels are displayed in reverse to read properly as +5 V and
Gnd when looking at them from the track side.

2.5 The PIC12F629 – Flashing LED Application

The first simple program developed in this section turns ON and OFF the
LED connected to the GP1 (general purpose I/O 1) pin. The LED is to be
turned ON for about 0.5 s and then turned OFF for another 0.5 s. This
process is then repeated continuously. Exact timing is not of concern
for this program. It will be explained in the next chapter of the book,
where timers available in the PIC microcontroller family are discussed in
more detail. Both C and assembler versions of the same program will be
considered in this section.

2.5.1 The PIC12F629 – Flashing LED – C Program

To compile C programs in this chapter we have used the 'PICC Lite V9.6'
compiler from HI-TECH, which is free evaluation software with a code
size limit. Again, this limited but free piece of software is adequate for all
applications discussed in this book.

The C program listed in Listing 2.1 contains a main function (`main()`)
and a delay function (`delay()`) to produce the required time delay
between turning the LED ON and OFF. In the main part of the program
the LED is first turned ON and the function `delay()` is called. After a
time delay the LED is turned OFF and the function `delay()` is called

again before the program returns to the beginning of the while(1) loop where the LED is turned ON again. This is repeated continuously to give the flashing LED.

```
1   #include <pic12f6x.h>
2   void delay(void) {           // 450 mS delay
3     int j;                      // declare a variable
4     for(j=0;j<30000;j++);       // produce a delay
5   }
6   void main(void)              // main function
7   {                            // begin of the program
8     TRIS1 = 0;                 // set port pin 1 to output
9     while (1){                 // repeat for ever
10    GPIO1= 1;                  // port pin = 1
11    delay();                   // short delay
12    GPIO1= 0;                  // port pin = 0
13    delay();                   // short delay
14    }          // end of repeat for ever
15  }            // end of the program
```

Listing 2.1   PIC12F629 flashing LED – C program

All programs shown in this book have line numbers for ease of reference. The HI-TECH C compiler generates these numbers for the ease of debugging but C language does NOT require the line numbers. To make programs more readable, a large number of comments are used for the programs discussed in this chapter. In C language we can include a comment using // (for a one-line comment) or /* . . . */ (for more lines of comments).

Line 1 of the program includes the file <pic12f6x.h> where labels such as GP1 are assigned to port addresses on the chip. This allows the use of labels through the program instead of numbers when referring to ports or registers of the PIC microcontroller, which makes programs more readable and easier to follow. Lines 2 to 6 make up the delay() function, which, when called by the main() function, goes through a for loop counting from 0 to 30,000 and then returns to the main() function. This counting takes approximately 0.5 s of the program execution time on the PIC12F629 microcontroller. Line 7 is the start of the main() function, which is, like every function in C, enclosed in the curly brackets, {}, between lines 8 and 13. This program needs to flash the LED continuously, so switching the LED ON/OFF is placed inside a while(1){ . . . } loop between lines 10 and 13. These statements will therefore execute infinitely, or until the program is interrupted externally (for example, if the power supply is switched off). The LED is turned ON at line 10 and then the delay function is called at line 11 to generate a delay of about 0.5 s. The LED is then turned OFF at line 12 followed by a delay at line 13.

The port pins in PIC need to be configured as inputs or outputs. To configure GP1 as an output pin, a 0 should be written into a corresponding configuration bit (TRIS1), which is done in this program in line 8.

2.5.2 PIC12F629 – Flashing LED – Assembler Program

To write assembler programs in this book we have used 'MPLAB IDE 7.5', the development tool freely available from Microchip. This free version of the assembler tool, like the C compiler from HI-TECH, has a code size limit, but this was not a problem for the programs developed for this book.

A delay subroutine to produce the time delay required between turning the LED ON and OFF starts at line 21 (Listing 2.2) and extends to line 32. The LED is turned ON at line 14 and the delay subroutine is called on the following line. After a time delay the LED is turned OFF and the delay subroutine is called again. This is repeated continuously, which results in a flashing LED.

Like the C program discussed earlier, the assembler program generates line numbers for ease of debugging. Those line numbers are shown on Listing 2.2 as well for ease of reference in this description of the program operation. Comments are also provided in order for the reader to follow and understand the program easily. In assembler language program comments need to be preceeded by semicolons ';'. Line 1, for example, shows a comment proceeded by ';' to explain the purpose of the program. In line 2, we have included a file called p12f629.inc so that we can refer to ports and registers inside the PIC by names (for example, STATUS) rather than numbers (such as 03H). Lines 5, 6 and 7 assign three registers to three labels T1, T2 and T3 to be used in the delay subroutine.

```
1   ;flashingLED12F629.asm
2   include "p12f629.inc"
3
4   ; global variables
5   T1      equ     26h
6   T2      equ     27h
7   T3      equ     28h
8
9   Main    bsf     STATUS,5        ; select bank 1 for TRISIO
10          movlw   B'11111101'     ; GP1 as output
11          movwf   TRISIO          ; set GP1 as output
12          bcf     STATUS,5        ; select bank 0
```

Listing 2.2 PIC12F629 flashing LED – assembly program

```
13
14   again    bsf     GPIO,GP1      ; port pin = 1
15            call    Delay         ; 0.5 second delay
16            bcf     GPIO,GP1      ; port pin = 1
17            call    Delay         ; 0.5 second delay
18            goto    again         ; repeat
19
20   ; Delay 0.5 sec
21   Delay    movlw   3             ; load values
22            movwf   T1            ; T1 = 3
23            movlw   0EBh          ; W = 235
24   DelT1    movwf   T2            ; T2 = 235
25   DelT2    movwf   T3            ; T3 = 235
26   DelT3    decfsz  T3,1          ; T3 = T3-1, repeat not 0
27            goto    DelT3         ; repeat previous line
28            decfsz  T2,1          ; T2 = T2-1, repeat not 0
29            goto    DelT2         ; repeat previous 4 lines
30            decfsz  T1,1          ; T3 = T3-1, repeat not 0
31            goto    DelT1         ; repeat previous 6 lines
32            return
33       ;
34   end
```

Listing 2.2 (*continued*)

As mentioned in the previous C programming section, port pins in PIC can be configured as inputs or outputs. To configure the GP1 pin as an output pin a 0 should be written into the corresponding bit inside TRISIO. To write to the TRISIO register we need to be able to access the appropriate register bank, which is register bank 1 (Figure 2.6). Register bank 1 is selected by setting bit 5 (RP0) in the status register. This is done in line 9 of the program by using the bsf (Bit Set File register) instruction to set bit 5 of the status register (Figure 2.7).

	File Address		File Address
Indirect addr. [1]	00h	Indirect addr. [1]	80h
TMRO	01h	OPTION_REG	81h
PCL	02h	PCL	82h
STATUS	03h	STATUS	83h
FSR	04h	FSR	84h
GPIO	05h	TRISIO	85h

Figure 2.6 Part of PIC12F629 DATA memory

© Microchip Technology Inc. Reproduced with permission.

Reserved	Reserved	R/W-0	R-1	R-1	R/W-x	R/W-x	R/W-x
IRP	RP1	RP0	$\overline{\text{TO}}$	$\overline{\text{PD}}$	Z	DC	C

bit 7   bit 0

Figure 2.7 Status register (address 03h or 83h)

© Microchip Technology Inc. Reproduced with permission.

To configure GP1 as an output pin, we first load the W register (line 10) with a binary value of '11111101' and then we load this value into the TRISIO register (line 11). Note that, in this way, bit 1 of the TRISIO register is set to 0 and GP1 is therefore configured as an output pin.

In order to access the GP register for I/O operations we use the `bcf` (Bit Clear File register) instruction to clear bits of the status register to return to register bank 0 (line 12).

The aim of this program is to flash the LED continuously and code for switching the LED ON and OFF is placed between lines 14 and 18. The LED is turned ON at line 14 using the `bsf` instruction and then the delay of about 0.5s is called at line 15. The LED is then turned OFF at line 16 using the `bcf` instruction again, followed by the call to the delay subroutine at line 17. A `goto` instruction is used at line 18 to send the flow of the program to line 14 again.

2.5.3 PIC12F629 – Flashing LED with a Push Button

The program developed in this section has the function of turning the LED connected to GP1 ON and OFF only when the push button connected to GP4 is pressed. This flashing of the LED is going to be repeated as long as the push button is pressed and the flashing stops when it is released.

2.5.3.1 PIC12F629 – Flashing LED with a push button – C program

```
1  #include <pic12f6x.h>
2
3  void delay(void) {        // 450 mS delay
4    int j;                    // declare a variable
5    for(j=0;j<30000;j++)    // produce a delay
6      ;
7  }
```

Listing 2.3 PIC12F629 flashing LED with a push button – C program

```
 8  void main(void){
 9    TRIS1 = 0;              // set port pin 1 to output
10    TRIS4 = 1;              // set port pin 4 to input
11
12    while (1){              // repeat for ever
13      if(GPIO4==1){         // is push button pressed?
14        GPIO1 = 1;          // port pin = 1
15        delay();            // short delay
16        GPIO1 = 0;          // port pin = 0
17        delay();            // short delay
18      }   // end of if
19    }     // end of repeat for ever
20  }
```

Listing 2.3   (*continued*)

This program is very similar to the previous program shown in Listing 2.1 with a few modifications explained below.

At line 10, GP4 is configured as an input by setting the relevant bit in the TRIS register. At line 13 an `if` statement is added to check the logic on the GP4 input pin. If the button is pressed, the logic on this pin will be 1 and lines 14 to 17 between curly brackets will be executed. The LED will be switched ON and OFF and this sequence will be repeated as long as the push button is pressed, i.e. as long as the logic 1 is present on the pin GP4. No action is taken otherwise and the program just keeps checking the state of the pin GP4 on line 13.

2.5.3.2 PIC12F629 – Flashing LED with a push button – assembly program

```
 1  ; SW_LED12F629.asm
 2  include "p12f629.inc"
 3
 4  ; global variables
 5  T1     equ     26h
 6  T2     equ     27h
 7  T3     equ     28h
 8
 9  Main   bsf     STATUS,5    ; select bank 1 for TRISIO
10         movlw   B'11111101' ; GP1 as output, rest inputs
11         movwf   TRISIO      ; set GP1 as output
12         bcf     STATUS,5    ; select bank 0
```

Listing 2.4   PIC12F629 flashing LED with a push
button – assembly program

```
13
14   again  btfss  GPIO,GP4    ; is push button pressed?
15          goto   again       ; not pressed, read again
16          bsf    GPIO,GP1    ; port pin = 1
17          call   Delay       ; 0.5 second delay
18          bcf    GPIO,GP1    ; port pin = 1
19          call   Delay       ; 0.5 delay
20          goto   again       ; repeat
21
22   ; Delay 0.5 sec
23   Delay movlw  3            ; load values
24         movwf  T1           ; T1 = 3
25         movlw  0EBh         ; W = 235
26   DelT1 movwf  T2           ; T2 = 235
27   DelT2 movwf  T3           ; T3 = 235
28   DelT3 decfsz T3,1         ; T3 = T3-1, rpeat not 0
29         goto   DelT3        ; repeat previous line
30         decfsz T2,1         ; T2 = T2-1, rpeat not 0
31         goto   DelT2        ; repeat previous 4 lines
32         decfsz T1,1         ; T3 = T3-1, rpeat not 0
33         goto   DelT1        ; repeat previous 6 lines
34         return
35         ;
36   end
```

Listing 2.4 (*continued*)

The modification introduced in the assembler program (compared with the previous assembler program in Listing 2.2) is to configure GP4 as an input pin, which is done at line 10. The W register is loaded with the constant that sets bit 5 (GP4 bit of the STATUS register) and hence configures GP4 an input. The other modifications are made at lines 14 and 15, where the instruction btfss (Bit Test File register Skip if Set) at line 14 checks GP4 input bit and, if it is 1 (push button pressed), skips the next instruction goto at line 15, and executes the rest of the program starting at line 16. If GP4 is 0 (push button released) then the goto instruction at line 15 is executed and the program goes back to line 14 again to check the state of the GP4 pin.

2.5.4 PIC12F629 – Two Flashing LEDs and a Push Button

The program in this section turns the LED connected to GP0 ON and OFF when the push button is pressed. If the push button is released or is not pressed, the LED connected to GP1 is turned ON and OFF instead.

2.5.4.1 PIC12F629 – Two flashing LEDs and a push button – C program

```
1   #include <pic12f6x.h>
2
3   void delay(void) {        // 450 mS delay
4     int j;                      // declare a variable
5     for(j=0;j<30000;j++)    // produce a delay
6       ;
7   }
8
9   void main(void){
10    TRIS0 = 0;              // set port pin 0 to output
11    TRIS1 = 0;              // set port pin 1 to output
12    TRIS4 = 1;              // set port pin 4 to input
13
14    while (1){              // repeat for ever
15      if(GPIO4==1){         // is push button pressed?
16        GPIO0 = 1;          // port pin = 1
17        delay();            // short delay
18        GPIO0 = 0;          // port pin = 0
19        delay();            // short delay
20      }     // end of if
21      else{                 // otherwise
22        GPIO1 = 1;          // port pin = 1
23        delay();            // short delay
24        GPIO1 = 0;          // port pin = 0
25        delay();            // short delay
26      }     // end of else
27    }                       // end of repeat for ever
28  }
```

Listing 2.5   PIC12F629 with two flashing LEDs and a push button – C program

The program required to achieve this task is given in Listing 2.5. This program is, of course, very similar to the one shown in Listing 2.3, with some important modifications. An extra output pin – GP0 – is needed in this program. Line 10 takes care of configuring the GP0 as an output pin. The if statement in line 15 and the else statement at line 21 are used to check if the push button is pressed, in other words if GP4 is set to logic 1. If the push button is pressed, lines 16 to 19 are executed causing the LED connected to GP0 to flash. If the push button is not pressed, the lines following the else statement (lines 22–25) are executed and this results in flashing the LED connected to the GP1 pin.

2.5.4.2 PIC12F629 – Two flashing LEDs and a push button – assembly program

```
1    ; SW_LED12F629.asm
2    include      "p12f629.inc"
3
4    ; global variables
5    T1        equ      26h
6    T2        equ      27h
7    T3        equ      28h
8
9    Main bsf       STATUS,5    ; select bank 1 for TRISIO
10        movlw     B'11111100' ; GP0,GP1 outputs, rest inputs
11        movwf     TRISIO      ; set GP1 as output
12        bcf       STATUS,5    ; select bank 0
13
14   again btfss GPIO,GP4       ; is push button pressed?
15        goto   NOTkey         ; not pressed, read again
16        bsf    GPIO,GP0       ; GP0 = 1
17        call   Delay          ; 0.5 second delay
18        bcf    GPIO,GP0       ; GP0 = 0
19        call   Delay          ; 0.5 second delay
20        goto   again          ; repeat
21   NOTkey bsf     GPIO,GP1    ; GP1 = 1
22        call    Delay         ; 0.5 second delay
23        bcf     GPIO,GP1      ; GP1 = 0
24        call    Delay         ; 0.5 second delay
25        goto    again         ; repeat
26
27   ; Delay 0.5 sec
28   Delay movlw 3              ; load values
29         movwf T1             ; T1 = 3
30         movlw 0EBh           ; W = 235
31   DelT1 movwf T2             ; T2 = 235
32   DelT2 movwf T3             ; T3 = 235
33   DelT3 decfsz T3,1          ; T3 = T3-1, rpeat not 0
34         goto DelT3           ; repeat previous line
35         decfsz T2,1          ; T2 = T2-1, rpeat not 0
36         goto DelT2           ; repeat previous 4 lines
37         decfsz T1,1          ; T3 = T3-1, rpeat not 0
38         goto DelT1           ; repeat previous 6 lines
39         return
40         ;
41   end
```

Listing 2.6   PIC12F629 with two flashing LEDs and
a push button – assembly program

Here, according to the program in Listing 2.6, we are required to configure GP0 as an output pin, which is done at line 10 with bit 0 being cleared first in the W register and then copied over to TRISIO. Most of the program is similar to the previous assembler program listed in Listing 2.4, with some modifications. To implement the required program flow, the push-button press is tested at line 14. If the button is not pressed (logic 0 on the GP0 pin) the `goto` instruction at line 15 is executed and the program continues to line 21 where the LED connected to GP1 is flashed. If the push button is pressed, line 15 is ignored and the LED connected to GP0 is flashed (lines 16–20). After that, the program flow is returned to line 14 where the button press is tested again.

2.6 PIC16F627A Circuit Design

The remainder of the programs presented in this chapter make use of a circuit based around the PIC16F627A microcontroller shown in Figure 2.8. Two push buttons are connected to RA4 and RA5 pins, which are configured as inputs. A common cathode seven-segment display is connected to port RB. The push buttons are active low – when pressed the input pins are at logic 0 and when released the pins are at logic 1. To switch on a segment of the common cathode display, we have to apply logic 1 to that segment.

Referring to Figure 2.8, the pins RA4 and RA5 are normally connected to a 5 V supply voltage through R9 and R10 resistors and are

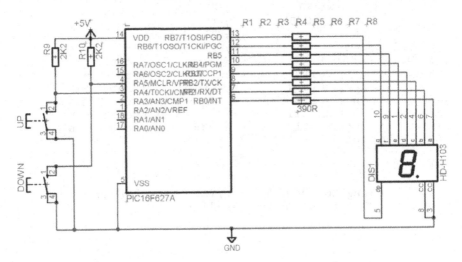

Figure 2.8 PIC16F627A microcontroller schematic

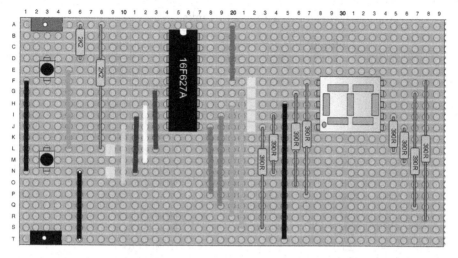

Figure 2.9 PIC 16F627A microcontroller strip board design

read as logic 1s by the program inside the PIC. When the push buttons are pressed, RA4 and RA5 pins are connected the 0 V supply voltage, so the input would be read as logic 0 by the program inside the PIC.

2.7 PIC16F629 Strip Board Design

Here, for ease of constructing a board, we have used a strip board with 20 rows and 39 columns (see Figure 2.9). Rows A and T are assigned to a +5 V track and to a 0 V track respectively. The PIC12F629 is placed on rows B to J, and columns 14 to 17. Tracks B to J on the column 15 are cut to make sure there is no connection between the left and right sides of the chip (pin 1 is NOT connected to pin 18). Tracks L and N on column 9, tracks F, G, H, I and J on column 22 are also cut. The other links used to construct the circuit are listed in Table 2.1.

2.7.1 Some Notes on Seven-Segment Display

The segments of a seven-segment display are shown in Figure 2.10. There are two types of these displays – common anode and common cathode. In common cathode displays all the cathodes are connected together and in common anode the anodes are connected. In common

Table 2.1 Links

Column	Connection
1	from F to N
5	from E to L
6	from N to T
10	from J to O
11	from I to N
12	from H to M
13	from G to L
18	from J to P
19	from I to Q
20	from A to F
20	from H to R
21	from G to S
25	from H to T

cathode displays the cathode is connected to the ground and application of voltage to the individual segments would light the segments. In common anode, the anode is connected to the supply and segments are lit by connecting them to the ground. Here we have used a common cathode display.

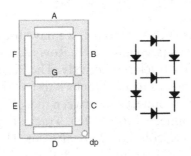

Figure 2.10 Seven-segment display

2.8 PIC16F627A PCB Board Design

Figure 2.11 shows the PCB board for the circuit designed and shown in Figure 2.8.

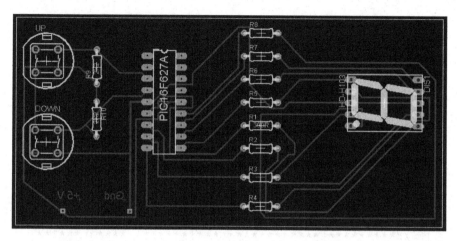

Figure 2.11 PIC 16F627A microcontroller PCB board design

2.9 PIC16F627A – Display Segments

The first program developed in this section will display each segment of the seven-segment display in turn for periods of about 0.5 s. The second program will use the seven-segment display to display the numbers from 0 to 9. Finally, the third program will count up and display the count if the first push button is pressed and count down and again display the count if the second push button is pressed.

2.9.1 PIC16F627A – Display Segments – C Program

A C program shown in Listing 2.7 is written in order to light the segments of the seven-segment display (a, b, c, d, e, f, g and dp – decimal point) one at a time for half a second. Lines 1, 2, 3, and 4 are comments to show how the hardware is connected: segment A (a) is connected to the RB0 pin, segment B (b) is connected to RB1 and so on. For a segment to be displayed, we have declared a variable of type integer (line 10) called seg. To configure all bits of port B as outputs, we have set TRISB to 0x00 (hexadecimal number 0). All bits are set to zeros, hence all pins are outputs. To switch on the appropriate segment, we use a switch(. . .) case{ . . . } C statement. This works as follows. Referring to line 13, if the value of seg in the switch brackets is 0, then line 14 – case 0 – is executed. If seg is 1, line 17 is executed and so on. When a case statement is selected to be executed, all statements following the case would be executed, until a break statement is reached. So when case 0 is executed, PORTB is set to 0b00000001, which in turn switches on segment a of the display. Delay is called (line 15) and, since the break statement follows, the program breaks away from the switch(. . .) case { . . . } loop and continues execution at line 39. Here, the value of variable seg is incremented, in order to light up the next segment of the display after going through the switch(. . .) case{ . . . } loop again. Line 40 checks to make sure the value of seg is not bigger than 7. Can you understand why this check is necessary?

```
1   //*********************************************
2   // RB         7    6    5    4    3    2    1    0 *
3   // Segments  dp    G    F    E    D    C    B    A *
4   //*********************************************
5   #include <pic16f62xa.h>
6
7   void delay(void);
8
9   void main(void){
10    int seg=0;                  // variable to choose segments
11    TRISB = 0x00;               // make port B, all bits, outputs
12    while (1){                  // run for ever
13      switch(seg) {
14      case 0: PORTB = 0b00000001;              // A is ON
15        delay();
16        break;
17      case 1: PORTB = 0b00000010;              // B is ON
18        delay();
19        break;
20      case 2: PORTB = 0b00000100;              // C is ON
21        delay();
22        break;
23      case 3: PORTB = 0b00001000;              // D is ON
24        delay();
25        break;
26      case 4: PORTB = 0b00010000;              // E is ON
27        delay();
28        break;
29      case 5: PORTB = 0b00100000;              // F is ON
30        delay();
31        break;
32      case 6: PORTB = 0b01000000;              // G is ON
33        delay();
```

Listing 2.7 PIC16F627A display segments – C program

```
34        break;
35        case 7: PORTB = 0b10000000;                // dp is ON
36          delay();
37          break;
38        }
39        seg = seg + 1;
40        if(seg == 8) seg=0;
41      }
42    }
43
44    void delay(void) {        // 450 mS delay
45      int j;                        // declare a variable
46      for(j=0;j<30000;j++)  // produce a delay
47        ;
48    }
```

Listing 2.7 *(continued)*

2.9.2 PIC16F627A – Display Segments – Assembly Program

The assembly version of the previous program is shown in Listing 2.8. Lines 13 and 14 are used to configure all eight bits of PORTB as output bits. Lines 16 and 17 switch on segment A and a delay subroutine is called after that from line 18. Writing hexadecimal number 0x01 (binary 0000 0001) would turn on segment A of the display as it is connected to bit 0 of PORT B (as shown in the header of the program listing). 0x02 (0000 0010) sent to PORTB would light segment B and so on. Segment B is turned ON at lines 19 and 20 and the rest of the segments at lines (22, 23), (25, 26), (28, 29), (31, 32), (34, 35) and finally dp at (37, 38).

```
1    //***********************************************
2    // RB          7    6    5    4    3    2    1    0 *
3    // Segments   dp    G    F    E    D    C    B    A *
4    //***********************************************
5
6    include "p16f627A.inc"
7    ; global variables
8    T1      equ                    26h
9    T2      equ                    27h
10   T3      equ                    28h
```

Listing 2.8 PIC16F627A display segments – assembly program

```
11
12   ;****Set up the port****
13        movlw 0x00            ; Set the Port B pins
14        tris  PORTB           ; to outputs
15   ;
16   Start movlw 0x01
17        movwf PORTB           ; A is ON
18        call  Delay           ; time Delay
19        movlw 0x02
20        movwf PORTB           ; B is ON
21        call  Delay           ; time Delay
22        movlw 0x04
23        movwf PORTB           ; C is ON
24        call  Delay           ; time Delay
25        movlw 0x08
26        movwf PORTB           ; D is ON
27        call  Delay           ; time Delay
28        movlw 0x10
29        movwf PORTB           ; E is ON
30        call  Delay           ; time Delay
31        movlw 0x20
32        movwf PORTB           ; F is ON
33        call  Delay           ; time Delay
34        movlw 0x40
35        movwf PORTB           ; G is ON
36        call  Delay           ; time Delay
37        movlw 0x80
38        movwf PORTB           ; dp is ON
39        call  Delay           ; time Delay
40
41        goto Start            ; go back to Start
42
43   ; Delay 0.5 sec
44   Delay movlw 3                   ; load values
45        movwf  T1               ; T1 = 3
46        movlw  0EBh             ; W = 235
47   DelT1  movwf   T2             ; T2 = 235
48   DelT2  movwf   T3             ; T3 = 235
49   DelT3  decfsz T3,1            ; T3 = T3-1, rpeat not 0
50        goto    DelT3         ; repeat previous line
51        decfsz T2,1            ; T2 = T2-1, rpeat not 0
52        goto    DelT2         ; repeat previous 4 lines
53        decfsz T1,1            ; T3 = T3-1, rpeat not 0
54        goto    DelT1         ; repeat previous 6 lines
55        return                  ; return from delay
56   ;
57   end
```

Listing 2.8   (*continued*)

2.9.3 PIC16F627A – Displaying Numbers – C Program

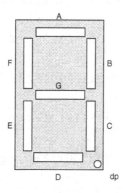

Listing 2.9 gives a listing of the program that displays the numbers 0 to 9 one at a time for half a second on the seven-segment display. To display the number 0, for example, we need to switch on the segments A, B, C, D, E and F. In line 14 of the program bits connected to those segments are set. To display the number 1 we only need to enable segments B and C (line 17). To display number 2 we enable segments A, B, D, E and G (line 20). We carry on in the same way for the rest of the numbers. As in previous examples this process is repeated continuously – the program runs for ever or until it is externally terminated.

```
1    //*********************************************
2    // This program writes numbers 0 to 9 to the display *
3    // Segments      dp    G    F    E    D    C    B    A*
4    //*********************************************
5    #include <pic16f62xa.h>
6
7    void delay(void);
8
9    void main(void){
10     int seg=0;                  // variable to choose segments
11     TRISB = 0x00;               // make port B, all bits, outputs
12     while (1){                  // run for ever
13       switch(seg) {
14         case 0: PORTB = 0b00111111;// display 0
15         delay();
16       break;
17         case 1: PORTB = 0b00000110;// display 1
18         delay();
19       break;
20         case 2: PORTB = 0b01011011;// display 2
21         delay();
```

Listing 2.9   PIC16F627A display numbers – C program

```
22      break;
23      case 3: PORTB = 0b01001111;// display 3
24        delay();
25      break;
26      case 4: PORTB = 0b01100110;// display 4
27        delay();
28      break;
29      case 5: PORTB = 0b01101101;// display 5
30        delay();
31      break;
32      case 6: PORTB = 0b01111101;// display 6
33        delay();
34      break;
35      case 7: PORTB = 0b00000111;// display 7
36        delay();
37      break;
38      case 8: PORTB = 0b01111111;// display 8
39        delay();
40      break;
41      case 9: PORTB = 0b01101111;// display 9
42        delay();
43      break;
44      }
45      seg = seg + 1;
46      if(seg == 10) seg=0;
47      }
48   }
49   void delay(void) {    // 450 mS delay
50      int j;                    // declare a variable
51        for(j=0;j<30000;j++)    // produce a delay
52          ;
53   }
```

Listing 2.9   (*continued*)

2.9.4 PIC16F627A – Displaying Numbers – Assembly Program

An assembly-language program to display numbers from 0 to 9 one at a time on the seven-segment display for half a second is shown in Listing 2.10. As in the C program to display number 0, we need to switch on segments A, B, C, D, E, and F. So in lines 16 and 17 we enable the appropriate bits. To display number 1 we only need to enable segments B and C, which we do in lines 19 and 20. To display number 2 we enable segments A, B, D, E and G in lines 22 and 23. Subroutine Delay is called after each number is displayed to make it visible to a human observer. The same procedure is carried out for the rest of the numbers.

```
 1   ;**********************************************
 2   ; This program writes numbers 0 to 9 to the display *
 3   ; Segments     dp   G   F   E   D   C   B  A *
 4   ;**********************************************
 5
 6   include "p16f627A.inc"
 7   ; global variables
 8   T1      equ                26h
 9   T2      equ                27h
10   T3      equ                28h
11
12   ;****Set up the port****
13      movlw  0x00             ; Set the Port B pins
14      tris   PORTB            ; to outputs
15   ;
16   Start movlw 0x3F           ; 00111111 : 3F : 0
17      movwf  PORTB
18      call                    Delay; Wait for some time
19      movlw  0x06             ; 00000110 : 06 : 1
20      movwf  PORTB
21      call                    Delay; Wait for some time
22      movlw  0x5B             ; 01011011 : 5B : 2
23      movwf  PORTB
24      call                    Delay; Wait for some time
25      movlw  0x4F             ; 01001111 : 4F : 3
26      movwf  PORTB
27      call                    Delay; Wait for some time
28      movlw  0x66             ; 01100110 : 66 : 4
29      movwf  PORTB
30      call                    Delay; Wait for some time
31
32      movlw  0x6D             ; 01101101 : 6D : 5
33      movwf  PORTB
34      call                    Delay; Wait for some time
35      movlw  0x7D             ; 01111101 : 7D : 6
36      movwf  PORTB
37      call                    Delay; Wait for some time
38      movlw  0x07             ; 00000111 : 3F : 7
39      movwf  PORTB
40      call                    Delay; Wait for some time
41      movlw  0x7F             ; 01111111 : 7F : 8
42      movwf  PORTB
43      call                    Delay; Wait for some time
44      movlw  0x67             ; 01100111 : 67 : 9
45      movwf  PORTB
46      call                    Delay; Wait for some time
47      goto                    Start; go back to Start
48
49   ; Delay 0.5 sec
50   Delay movlw 3              ; load values
51      movwf  T1               ; T1 = 3
```

Listing 2.10   PIC16F627A display numbers – assembly program

```
52      movlw    OEBh          ; W = 235
53  DelT1 movwf     T2              ; T2 = 235
54  DelT2 movwf     T3              ; T3 = 235
55  DelT3 decfsz    T3,1          ; T3 = T3-1, rpeat not 0
56      goto                      DelT3; repeat previous line
57      decfsz T2,1          ; T2 = T2-1, rpeat not 0
58      goto                      DelT2; repeat previous 4 lines
59      decfsz T1,1          ; T3 = T3-1, rpeat not 0
60      goto DelT1            ; repeat previous 6 lines
61      return                ; return from delay
62      ;
63  end
```

Listing 2.10 (*continued*)

2.9.5 PIC16F627A – Up/Down Counter with Display – C Program

Listing 2.11 shows a C program that reads the state of two push buttons connected to RA4 and RA5 pins of the PIC, updates the number/count if any of those two buttons is pressed and displays that number on a seven-segment display. If the button connected to RA4 is pressed then the number to be displayed is incremented; if the other button connected to RA5 is pressed the number is decremented and displayed. The buttons are active low – pins connected to those buttons (RA4 and RA5) are at logic 1 if the buttons are not pressed and at logic 0 if they are pressed.

Line 12 of the program configures PORTB as an output port (all eight bits) and line 13 configures PORTA as input port. Line 15 checks if the switch is pressed; if it is, the num variable is incremented (num++, i.e. num=num+1). Line 16 checks the other switch and if it is pressed it decrements the num variable (num--, i.e. num=num-1). Line 17 checks to make sure that num has not exceeded 9 and if it has then this simple counter is reset to 0. Line 18 makes sure the number is not less than 0 – if it is, the number is set to 9.

```
1  //************************************************
2  // This program reads switches connected to RA4 and RA5 *
3  // and updates the displayed number accordingly by
   incrementing or decrementing *
4  //************************************************
5
6  #include <pic16f62xa.h>
7
8  void delay(void);
```

Listing 2.11 PIC16F627A Up/down counter with display – C program

```
 9
10    void main(void){
11      int num=4;                    // variable to choose segments
12      TRISB = 0x00;                 // make port B, all bits, outputs
13      TRISA = 0xFF;                 // make port A, all bits, inputs
14      while (1){                    // run for ever
15        if(RA4 == 0) num++;  // if pressed, num = num+1
16        if(RA5 == 0) num--;  // if pressed num = num-1
17        if(num>9) num = 0;   // 0 if num = 10
18        if(num<0) num = 9;   // 9 if num = -1
19
20        switch(num) {
21        case 0: PORTB = 0b00111111;              // display 0
22          delay();
23          break;
24        case 1: PORTB = 0b00000110;              // display 1
25          delay();
26          break;
27        case 2: PORTB = 0b01011011;              // display 2
28          delay();
29          break;
30        case 3: PORTB = 0b01001111;              // display 3
31          delay();
32          break;
33        case 4: PORTB = 0b01100110;              // display 4
34          delay();
35          break;
36        case 5: PORTB = 0b01101101;              // display 5
37          delay();
38          break;
39        case 6: PORTB = 0b01111101;              // display 6
40          delay();
41          break;
42        case 7: PORTB = 0b00000111;              // display 7
43          delay();
44          break;
45        case 8: PORTB = 0b01111111;              // display 8
46          delay();
47          break;
48        case 9: PORTB = 0b01101111;              // display 9
49          delay();
50          break;
51        }
52      }
53    }
54    void delay(void) {              // 450 mS delay
55      int j;                        // declare a variable
56      for(j=0;j<30000;j++) // produce a delay
57        ;
58    }
```

Listing 2.11 (*continued*)

2.9.6 PIC16F627A – Up/Down Counter with Display – Assembly Program

Listing 2.12 shows the assembly-level version of the same program, which increments or decrements the number according to the state of two push buttons and displays the number on the seven-segment display attached to PORTB. If the number exceeds 9 or becomes negative it is reset to 0 or 9 and counting continues.

Lines 14 and 15 configure PORTB as output port and lines 16 and 17 set PORTA as an input port. Checking of the switch connected to RA4 is done at line 19 and, if pressed, Num is incremented at line 20. State of the push button connected to RA5 is checked at line 21 and, if pressed, Num is decremented at line 22. To check if the Num exceeded 9 – if it is equal to 10 – we apply the exclusive OR operation between the Num and number 10. If Num is equal to 10 the result of this operation is 0, which would be indicated in the STATUS register. This is done in lines 24 to 27. Similarly, to check if Num is equal to -1 we need to apply the exclusive OR operation between Num and the number -1. Note that the number 0xFF is used in the exclusive OR operation and not -1 according to the two's complement rule illustrated below.

```
 1 ;**********************************************
 2 ; This program reads switches connected to RA4 and RA5*
 3 ; and updates the displayed number accordingly by
   incrementing or decrementing *
 4 ;**********************************************
 5 include "P16f627A.inc"
 6 ; global variables
 7   T1        equ    0x26
 8   T2        equ    0x27
 9   T3        equ    0x28
10   Num       equ    0x29
11
12 ;****Set up the port****
13
14 Main  movlw 0x00      ; Set the Port B pins
15       tris  PORTB      ; to outputs.
16       movlw 0Xff       ; Set the Port A pins
17       tris  PORTA      ; to inputs.
18 ;
19 Start btfss PORTA,4    ; check push button pressed
20       incf  Num,1      ; if pressed, Num = Num + 1
21       btfss PORTA,5    ; check push button pressed
22       decf  Num,1      ; if pressed, Num = Num - 1
23
24       movlw 0x0A       ; test if Num = 10
```

Listing 2.12   PIC16F627A up/down counter with display – assembly program

```
25          xorwf Num,0         ;
26          btfsc STATUS, 2     ;
27      goto    Num_0           ; if yes, Num = 0
28
29          movlw 0xFF          ; test if Num = -1
30          xorwf Num,0         ;
31          btfsc STATUS, 2     ;
32      goto  Num_9             ; if yes, Num = 9
33
34      goto  Num2W             ; -1 < Num < 10
35
36 Num_0  clrw                  ; W = 0 later Num = 0
37          movwf Num           ; save W into Num
38      goto  Num2W
39
40 Num_9  movlw 0x09            ; W = 9 later Num = 9
41          movwf Num
42
43 Num2W  movf  Num,0           ; put Num into W
44          call  Convert       ; convert to Segments
45          movwf PORTB         ; Display Num
46          call  Delay         ; Wait for some time
47      goto  Start             ; go back to Start
48
49 Convert addwf PCL            ; W = W + PC
50          retlw 0x3F          ; 00111111 : 3F: 0
51          retlw 0x06          ; 00000110 : 06: 1
52          retlw 0x5B          ; 01011011 : 5B: 2
53          retlw 0x4F          ; 01001111 : 4F: 3
54          retlw 0x66          ; 01100110 : 66: 4
55          retlw 0x6D          ; 01101101 : 6D: 5
56          retlw 0x7D          ; 01111101 : 7D: 6
57          retlw 0x07          ; 00000111 : 3F: 7
58          retlw 0x7F          ; 01111111 : 7F: 8
59          retlw 0x67          ; 01100111 : 67: 9
60                                                    ;
61 ;Delay 0.5 sec
62 Delay movlw   3              ; load values
63          movwf   T1          ; T1 = 3
64          movlw   0EBh        ; W = 235
65 DelT1 movwf   T2             ; T2 = 235
66 DelT2 movwf   T3             ; T3 = 235
67 DelT3 decfsz T3,1            ; T3 = T3-1, rpeat not 0
68          goto                DelT3; repeat previous line
69          decfsz  T2,1        ; T2 = T2-1, rpeat not 0
70          goto    DelT2       ; repeat previous 4 lines
71          decfsz  T1,1        ; T3 = T3-1, rpeat not 0
72          goto DelT1          ; repeat previous 6 lines
73          return              ; return from delay
74                              ;
75  end
```

Listing 2.12 (*continued*)

Display Interfaces

3.1 Introduction

In this chapter, two display circuits based on the PIC16F627A micro-controller are developed and their operation is described in detail. In the first part of the chapter a PIC interface to a four-digit, seven-segment LED display is discussed and two programs using C and assembler are developed to make use of this circuit. In the second part we concentrate on interfacing the LCD to a PIC microcontroller and again present both C and assembler programs used to display a set of characters on the LCD display.

For the four-digit, seven-segment circuit, an HDSPB03E common cathode display, a resistor array, a 74HC4511 seven-segment decoder/driver and four NPN transistors with base resistors are used.

For the LCD display, a character-type dot matrix LCD module, Trident (JM162A) 16 × 2 line, back lit, HD44870 compatible and a 10K potentiometer are used.

Figure 3.1 shows MPLAB settings for programming the PICs in this chapter.

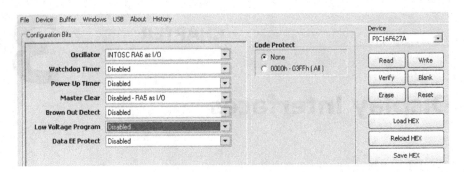

Figure 3.1 MPLAB programming setup

© Microchip Technology Inc. Reproduced with permission.

3.2 PIC16F627A Four-Digit, Seven-Segment LED Display Circuit Design

In this section we will describe the circuitry based on the PIC16F627A microcontroller and discuss both C and assembly versions of the program to count from 0000–9999 and display this count on a four-digit, seven-segment display. The developed circuit could be used for future applications and projects requiring four-digit, seven-segment displays including revolution and frequency counters, voltmeters and similar projects. Some of those projects will be described and implemented in the last chapter of the book, dealing with miscellaneous PIC projects. The schematic diagram of this circuit, drawn using the 'EAGLE 4.16 r2 Lite' software package, is shown in Figure 3.2.

We used PORTB pins as the input/output pins in this application while deliberately leaving PORTA pins unused in order to make use of them for applications where analog-to-digital conversions are required in addition to the display of data. The lower nibble (4 bits) of PORTB will be used for the value of the character to be displayed and is connected to the input pins of a 74HC4511 seven-segment decoder/driver. The upper nibble is connected to the NPN transistors to allow the switching (ON/OFF) of individual digits, one at a time. Current through each LED needs to be limited using additional resistors. To accomplish this we use a $220\,\Omega$ resistor array chip, which simplifies the circuit-building task.

3.3 PIC16F627A Four-Digit, Seven-Segment LED Display Circuit Strip Board Design

We have used a strip board to build the circuit from Figure 3.2. This strip board is shown in Figure 3.3. The PCB layout of the same circuit is

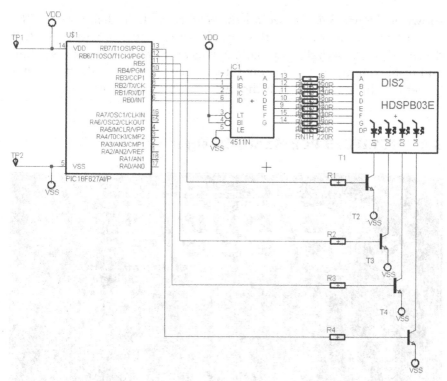

Figure 3.2 PIC16F627A four-digit, seven-segment LED display schematic

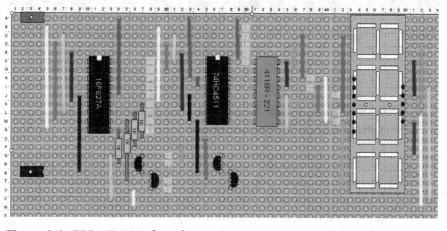

Figure 3.3 PIC16F627A four-digit, seven-segment LED display circuit strip
board design

shown in Figure 3.4. To see all the details of these diagrams for the readers who want to build this circuit using a strip board or a PCB design, we recommend downloading diagrams from this book's web site and displaying them on the computer screen or printing them out in a bigger size in order to see all the details in the diagrams.

3.4 PIC12F629 PCB Board Design

Figure 3.4 PIC16F627A four-digit, seven-segment LED display circuit PCB board design

3.5 PIC16F627A Four-Digit, Seven-Segment LED Display Circuit Application

The program developed in this section simply counts from 0000 to 9999 and displays the current count on four digits of the seven-segment display. To display the count, the program constantly scans the display, displaying just one digit at a time for a very short duration. During that time the other three digit displays are turned off. This approach enables us to use only four digital lines to send the required digit value from microcontroller to a display. The time period during which each digit is displayed is so short that to a human eye all four digits should be clearly visible all of the time.

3.5.1 PIC16F627A Four-Digit, Seven-Segment LED Display – C Program

The C program listed in Listing 3.1 contains the following functions:

- *main()* – main *function*
- *display(int d4, int d3, int d2, int d1)* – display function to display digits d1, d2, d3 and d4
- *delay()* – delay function to produce the required time delay between switching the digits ON and OFF.

Lines 1 to 6 are program comments, showing the main system connections. In this application only PORTB pins are used – pins 0 to 3 are connected to 74HC4511 display driver chip and pins 4 to 7 are connected to four transistors.

```
 1  //* * * * * * * * * * * * * * * * * * * * * * * * * * * *
 2  // RB      7      6      5      4      3      2      1      0   *
 3  //                Transistors            74HC4511             *
 4  //               --------------                               *
 5  //               |1 |2 |3 |4 |                                *
 6  //* * * * * * * * * * * * * * * * * * * * * * * * * * * *
 7  #include <pic16f62xa.h>
 8  void delay(void);
 9  void display(int d4,int d3, int d2, int d1);
10  void main(void){
11    int d4=0,d3=0,d2=0,d1=0;
12    TRISB = 0x00;              // port B outputs
13    while (1){                 // run for ever
14      display(d1,d2,d3,d4);
15        d4++;                  // d4 = d4 + 1
16      if(d4 == 10) {           // if d4 = 10 then
17        d4 = 0;                // d4 = 0 and
18                d3++;          // d3 = d3 + 1
19        }
20      if(d3 == 10) {           // if d3 = 10 then
21        d3 = 0;                // d3 = 0 and
22        d2++;                  // d2 = d2 + 1
```

Listing 3.1   PIC16F627A four-digit, seven-segment LED display – C program

```
23        }
24   if (d2 == 10) {          // if d2 = 10 then
25       d2 = 0;              // d2 = 0 and
26         d1++;              // d1 = d1 + 1
27       }
28   if (d1 == 10) d1 = 0;    // if d1= 10 and
29     }                      // d1 = 0
30   }
31   void display(int d4, int d3, int d2, int d1){
32        unsigned char numDisp;
33
34     numDisp = (unsigned char)d1 | 0x80;
35     PORTB   = numDisp;      // send to 4511
36     delay();               // digit time ON
37         PORTB   = 0;       // all digits OFF
38
39     numDisp = (unsigned char)d2 | 0x40;
40     PORTB   = numDisp;      // send to 4511
41     delay();               // digit time ON
42     PORTB   = 0;           // all digits OFF
43
44     numDisp = (unsigned char)d3 | 0x20;
45     PORTB   = numDisp;      // send to 4511
46     delay();               // digit time ON
47     PORTB   = 0;           // all digits OFF
48
49     numDisp = (unsigned char)d4 | 0x10;
50     PORTB   = numDisp;      // send to 4511
51     delay();               // digit time ON
52     PORTB   = 0;           // all digits OF
53   }
54   void delay(void) {       // some delay
55     int j;                 // declare a variable
56     for(j =0;j<500;j++);// produce a delay
57   }
```

Listing 3.1   (*continued*)

Lines 8 and 9 contain function prototypes to declare the functions to be used in the main function, before they are defined. Those functions are defined after the main function, which starts at line 10. Line 11 declares and initializes to 0 all the variables d1, d2, d3, and d4 used in the main function, with d1 related to the most significant and d4 to the least significant digit to be displayed on the seven-segment display. The pins of PORTB are all set as outputs on line 12, using the TRISB instruction, where '0' configures the relevant pin as the output pin and '0x' specifies hexadecimal number format; 0x00 is the hexadecimal equivalent of the binary 00000000 pattern and configures all the pins of PORTB as outputs. The while (1) block between lines 13 and 29 is the block that does most of the work in this program as it repeats continually while the program runs. The first line of this block (14) calls the display routine to display the four digits d1, d2, d3 and d4, one after the other. This routine will be discussed in more detail later in this section. The next line in the 'while' loop (15) increments the value of the least significant digit. If this value is 10, the value of d4 is reset to 0 and d3 is incremented (lines 16, 17 and 18). The same process of checking the values of other digits to be displayed and incrementing the next digit if this value is 10 is repeated between lines 20 to 28 for digits d3, d2 and d1. When the value of most significant digit d1 reaches 10, it is reset to 0 and the full counts rolls over to start from decimal 0000 again.

The `display` function at line 31 is declared as `void display (int d4, int d3, int d2, int d1)`, where void in front of the function indicates no return value and the parameters in the brackets indicate that this function requires four variables of the type integer. Line 32 declares a variable of type 'unsigned char' called `numDisp`. The unsigned character type variable is an 8-bit variable and is therefore suitable to be sent to 8-bit ports of the PIC microcontroller. The variable, `numDisp`, is used in this program for sending digit values to PORTB to be displayed on the seven-segment display. Line 34 (as well as lines 39, 44, and 49) sets up the digit to be displayed. Referring to line 34, the digit d1 first needs to be converted to 'unsigned char' type. To achieve this, a C programming technique called 'casting' is used. The C expression:

```
numDisp =(unsigned char) d1;
```

will, for example, convert the integer type variable d1 into unsigned character variable `numDisp` without changing the actual value of the variable d1.

Since only four lower bits of variable d1 are important in this application we combine those four bits of d1 with the 4-bit binary pattern required to switch on the corresponding digit of the display: 1000_2 (hex 8) sent to the upper four bits of PORTB will turn on transistor 1 and activate digit 1 of the display, 0100_2 (hex 4) will turn on transistor 2 and activate digit 2 and so forth. To combine the value of the digit to be displayed and the selection of the digit together in one byte we use the logical 'OR' operation. Line 34

```
numDisp = (unsigned char) d1|0x80;
```

would first change the type of variable d1 into an 8-bit unsigned character and then copy the four least significant bits of d1 into the lower nibble of `numDisp` and store the binary pattern of four bits, 1000_2 (hex 8), into the upper nibble. Sending this value to PORTB would therefore activate digit 1 of the seven-segment display and display the number from d1 on it at the same time. The other three digits will be turned off because of three zeros from the upper nibble: 1000_2.

The data are therefore sent to PORTB at line 35, and the `delay` function is called at line 36 to allow enough time for the digit to be seen by the human eye. Line 37 turns digit 1 OFF. For all other digits, this process is repeated, with the exception of the values copied into the upper nibble of PORTB used for switching on the appropriate transistors (0x40 for digit 2, 0x20 for digit 3 and 0x01 for digit 4). The `delay` function, defined between lines 54 to 57, just produces enough time for the display to be seen. Some experimenting with the maximum count of the `for` loop used in this function might be needed if the PIC is to operate at frequencies other than the 4 MHz used in this example.

3.5.2 PIC16F627A Four-Digit, Seven-Segment LED Display – Assembler Program

The assembler program given in Listing 3.2, which is written for a four-digit, seven-segment display, consists of four parts. The first part, lines 11 to 18, assigns the memory locations to variables to be used in the program. For example T1, T2 and T3 are used in the `delay` subroutine and D1, D2, D3 and D4 are used in the `display` subroutine and the rest of the program. The second part consists of the main routine from lines 24 to 63, which initializes the variables, calls the display and updates the digits D1, D2, D3 and D4. The third part is the `display` routine from lines 65 to 90, which scans the display and displays the digits one at a time with the appropriate delay defined in the `delay` routine from lines 92 to 108.

The first part of the program and all declarations are self-explanatory; we will try to explain the second part of the program. Lines 25 and 26 are used to configure all bits of PORTB as outputs as this port is used to drive the seven-segment decoder 74HC4511 (RB0–RB3) and to switch ON/OFF transistors (RB4–RB7) attached to each of the digits. Lines 27 to 31 reset the value of the variables. After the initialization the main program starts at line 33 and ends at line 63. The task of this section is to first call the `display` subroutine and then to update the digits. For example, at line 34 the `display` subroutine is called, which displays all the numbers in D1, D2, D3 and D4 and then, at lines 35 to 41, D4 and D3 are updated. The updating is done by incrementing D4 (the least significant digit) and placing it in the W register, so the XOR operation with number 10 could be carried out. Combining two identical numbers in the logical XOR operation will result in 0. If the result of the XOR operation in line 37 is zero, the value of D4 is 10, so D4 should be reset to 0. This is done in line 40. The value of D3 should also be incremented at this stage which is done in line 41. This is repeated for digits D3, D2 and D1 in the rest of the routine.

```
1   ;* * * * * * * * * * * * * * * * * * * * * * * * * * * * * * *
2   ; RB          7   6   5   4   3   2   1   0   *
3   ; Segments   D1  D2  D3  D4   A   B   C   D   *
4   ;           ----------------  ----------------  *
5   ;              Transistors         74HC4511     *
6   ;           --------------                       *
7   ;              |1 |2 |3 |4 |                     *
8   ;* * * * * * * * * * * * * * * * * * * * * * * * * * * * * * *
9   include "p16f627A.inc"
10  ; global variables
11  T1  EQU  0x26
12  T2  EQU  0x27
13  T3  EQU  0x28
14  i   EQU  0x29
15  D1  EQU  0x2A
16  D2  EQU  0x2B
17  D3  EQU  0x2C
18  D4  EQU  0x2D
19  ;* * * * * * * * * * * * * * * * * * * * * * * * * * * * * * *
20  org 0x0000              ; RESET vector
21  RST   goto   START
22
23  ;* * * * * * * * * * * * * * * * * * * * * * * * * * * * * * *
24  START
25        movlw   0x00       ; Set the Port pins
26     tris      PORTB       ; to outputs.
27       clrf    i           ; i = 0
28       clrf    D1          ; D1 = 0
29       clrf    D2          ; D2 = 0
30       clrf    D3          ; D3 = 0
31       clrf    D4          ; D4 = 0
32
33  AGAIN
34        call    display     ; display all digits
35        incf    D4          ; D4 = D4 + 1
36        movf    D4,0        ; W = D4
37        xorlw   0x0A        ; is D4 = 10?
38        btfss   STATUS,Z    ; if D4 not 10, jump
39        goto    D3Check     ; to D3Check
40        clrf    D4          ; D4 = 0
41        incf    D3          ; D3 = D3 + 1
42  D3Check
```

Listing 3.2   PIC16F627A four-digit, seven-segment LED display – assembly program

```
43        movf  D3,0          ; W = D3
44        xorlw 0x0A          ; is D3 = 10?
45        btfss STATUS,Z      ; if D3 not 10, jump
46        goto  D2Check       ; to D2Check
47        clrf  D3            ; D3 = 0
48        incf  D2            ; D2 = D2 + 1
49  D2Check
50        movf  D2,0          ; W = D2
51        xorlw 0x0A          ; is D2 = 10?
52        btfss STATUS,Z      ; if D2 not 10, jump
53        goto  D1Check       ; to D1Check
54        clrf  D2            ; D2 = 0
55        incf  D1            ; D1 = D1 + 1
56  D1Check
57        movf  D1,0          ; W = D1
58        xorlw 0x0A          ; is D1 = 10?
59        btfss STATUS,Z      ; if D1 not 10, jump
60        goto  skip          ; to skip
61        clrf  D1            ; D1 = 0
62  skip
63        goto  AGAIN
64  ;* * * * * * * * * * * * * * * * * * * * * * * * * * * *
65  display                   ; data in D1,D2,D3,D4
66        movf  D1,0          ; W = D1
67        iorlw 0x10          ; enable D1
68        movwf PORTB         ; display digit
69        call  DELAY         ; wait a bit
70        clrw                ; W = 0
71        movwf PORTB         ; all digits OFF
72        movf  D2,0          ; W = D2
73        iorlw 0x20          ; enable D2
74        movwf PORTB         ; display digit
75        call  DELAY         ; wait a bit
76        clrw                ; W = 0
77        movwf PORTB         ; all digits OFF
78        movf  D3,0          ; W = D3
79        iorlw 0x40          ; enable D3
80        movwf PORTB         ; display digit
81        call  DELAY         ; wait a bit
82        clrw                ; W = 0
83        movwf PORTB         ; all digits OFF
84        movf  D4,0          ; W = D4
85        iorlw 0x80          ; enable D4
86        movwf PORTB         ; display digit
87        call  DELAY         ; wait a bit
88        clrw                ; W = 0
89        movwf PORTB         ; all digits OFF
90        return
91  ;* * * * * * * * * * * * * * * * * * * * * * * * * * * *
92  DELAY
```

Listing 3.2 (*continued*)

```
 93        movlw  D'3'              ; 1 cycle
 94        movwf  T1                ; 1"
 95   DelT1
 96        movlw    D'20'           ; 1 "
 97         movwf   T2              ; 1 "
 98   DelT2
 99        movlw    D'33'           ; 1 "
100        movwf    T3              ; 1 "
101   DelT3
102        decfsz   T3              ; 1 "
103        goto     DelT3           ; 2 "
104        decfsz   T2              ; 1 ", repeat not 0
105        goto     DelT2           ; 2 ",
106        decfsz   T1              ; 1 ", repeat not 0
107        goto     DelT1           ; 2 ",
108        return                   ; 2 ",
109   ;* * * * * * * * * * * * * * * * * * * * * * * * * * * * * * * *
110   end
```

Listing 3.2   (*continued*)

The display routine at line 65 starts by loading the W register with the value of D1 at line 66. At line 67, the W register is ORed with hex value of 0x10, which sets bit 7 of the W register. The W register is sent to PORTB at line 68, which displays D1. A delay is called at line 69 for the time the digits stay ON. At lines 70 and 71 the digit is switched OFF again. Lines 72 to 89 repeat the above process for the rest of the digits.

The delay routine consists of three nested loops. The counts to be decremented to zero in those loops are defined in variables T1, T2 and T3. T1 is the starting count used in the outermost loop and T3 is used in the inner loop. As a comment for each line in this routine, the number of cycles per instruction in that line is given. For the internal clock of 4 MHz used here we can calculate that each instruction cycle takes 1 μs to be executed. The exact duration of the whole delay routine can now be easily calculated. This is left as an exercise for the keen reader.

3.6   PIC16F627A LCD Display Circuit Design

In this section two versions – C and assembly of the program to display text messages on the LCD display circuit using PIC16F627A and additional circuitry – are developed and described in detail. The schematic diagram of the circuit that can be used in any application where display of 2 lines of text (up to 16 characters each) is needed is shown in Figure 3.5.

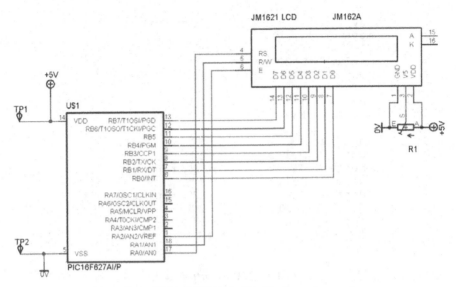

Figure 3.5 PIC16F627A LCD display schematic

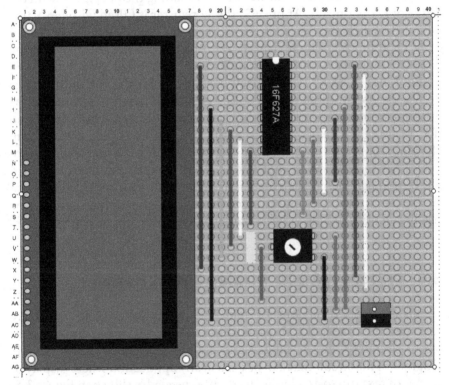

Figure 3.6 PIC16F627A LCD display circuit strip board design

As the LCD module requires eight data and three control lines, we use all pins on PORTB and three PORTA pins as the input/output pins in this application. The LCD is directly connected to the PIC micro and the only other component required for this circuit is a potentiometer, which is used to adjust the contrast of the LCD.

3.7   PIC16F627A Four-Digit, Seven-Segment LED Display Circuit Strip Board and PCB Design

Figure 3.6 shows the picture of the strip board and wiring we used for this application. The PCB schematic is shown in Figure 3.7. As with the seven-segment display application, this figure can be downloaded from

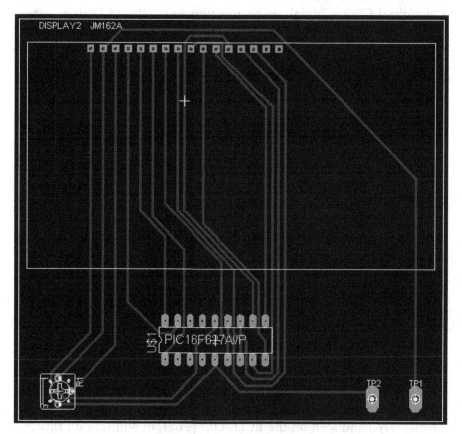

Figure 3.7 PIC16F627A LCD display circuit PCB board design

the book's web site in order to see the finer details and build a circuit correctly.

3.8 PIC16F627A LCD Display Circuit Application

The programs developed in this section display the first names of the authors but can be easily modified to display any combination of up to 32 characters at the time.

3.8.1 PIC16F627A LCD Display – C Program

The C program listed in Listing 3.3 is made up of the following functions:

- *main()* – main function
- *initializeLCD()* – initialization function to set the LCD to a certain mode of operation (for example 8-bit rather than the 4-bit interface)
- *sendData(unsigned char)* – function used to send the characters to be displayed to the LCD
- *sendCmd(unsigned char)* – function to send a command to the display for setting up things like display ON/OFF, cursor blinking and so forth
- *delay_mS(unsigned int)* – delay function used to produce varied required time delays for other functions.

Lines 1 to 7 are the initial program comments used to show the interface between the LCD and the PIC. Lines 10 to 18 are C directives used to assign labels to port pins and command numbers.

The main function of this program starts at line 60. The first statement at line 62 is an important and crucial statement for configuring PORTA pins in our system. This statement configures the PORTA pins (RA0, RA1 and RA2) as digital input/output pins, rather than analog input pins, a default configuration on PIC16F627A. Lines 63 and 64 configure pins of PORTB and PORTA as outputs. Lines 65, 66 and 67 clear the LCD control lines (RA0, RA1 and RA2) and at line 69 the initialization function discussed further in this section is called. Lines 70 to 85 are used to send the combination of the characters to be displayed on the LCD. Line 87 is a simple 'always true' while(1) statement. The purpose of this statement is to stop and hold the execution of the program on that line in order for the message to be displayed and seen on the LCD. Our program will stay on this line 'forever' or at least as long as the microcontroller is not reset.

```
 1    //* * * * * * * * * * * * * * * * * * * * * * * * * * * * * * * *
 2    // RB            7    6    5    4    3    2    1    0   *
 3    // Data Bus     D7   D6   D5   D4   D3   D2   D1   D0  *
 4    // ------------------------------------------------- *
 5    // RA            0    1    2                           *
 6    // Controls     RS   RW   E                            *
 7    //* * * * * * * * * * * * * * * * * * * * * * * * * * * * * * * *
 8    #include <pic16f62xa.h>
 9
10    #define RS                       RA0
11    #define RW                       RA1
12    #define E                        RA2
13    #define NewLine                  0xC0
14    #define ClrDisp                  0x01
15    #define TwoLine10dots            0x38
16    #define ScrOnCurOffBlinkOn       0x0F
17    #define IncCurDntMovDis          0x06
18    #define CurHome                  0x02
19
20    void delay_mS(unsigned int D)   {
21      unsigned int i,j;
22      for (i = 0; i < D; i++)
23      for (j = 0; j < 1000; j++)
24                   ;
25    }
26
27    void sendCmd(unsigned char Cmd){//write command to LCD
28      PORTB= Cmd;                    // put char into RAM
29      RS = 0;                        // command mode
30      RW = 0;                        // write operation
31      delay_mS(1);
32      E = 1;
33      delay_mS(1);
34      E = 0;                         // send command to LCD
35      return;
36    }
37
38    void sendData(unsigned char Ch){ // write text to LCD
39      PORTB= Ch;                     // put char into RAM
40      RS = 1;                        // data mode
41      RW = 0;                        // write operation
42      delay_mS(1);
43      E = 1;
44      delay_mS(1);
45      E = 0;                         // send data to LCD
46      return;
47    }
48
49    void   initializeLCD(void){
50          delay_mS(40);
51          sendCmd(TwoLine10dots);
```

Listing 3.3 PIC16F627A LCD display – C program

```
52              delay_mS(1);
53              sendCmd(ScrOnCurOffBlinkOn);
54              delay_mS(1);
55              sendCmd(ClrDisp);
56              delay_mS(1);
57              sendCmd(CurHome);
58      }
59
60      void main(void){
61
62              CMCON = 0x07;           // RA0,1,2 as I/Os
63              TRISA = 0x00;           // set to outputs
64              TRISB = 0x00;
65              RW = 0;                 // Write mode
66              RS = 0;                 // Command mode
67              E = 0;                  // Start talking to LCD
68
69              initializeLCD();
70              sendData('H');
71              sendData('a');
72              sendData('s');
73              sendData('s');
74              sendData('a');
75              sendData('n');
76              sendCmd(NewLine);
77              sendData('B');
78              sendData('r');
79              sendData('a');
80              sendData('n');
81              sendData('i');
82              sendData('s');
83              sendData('l');
84              sendData('a');
85              sendData('v');
86
87              while(1)                // wait for ever
88                      ;
89      }
```

Listing 3.3 (*continued*)

The LCD initialization function is called at line 49 from the main program. To achieve particular mode of operation the LCDs need to receive the certain commands with appropriate timing between those commands. Sequences of commands and timings vary from one manufacturer to another. For the LCD used in our system, the combination of commands and delays specified in the `initializeLCD()` function will set the LCD in the following way:

- two lines
- screen on

- cursor off
- blink on.

Line 55 clears the display and line 57 sends the blinking underscore to the beginning of the first line.

The sendCmd and sendData functions starting at lines 27 and 38 are exactly the same, except for lines 29 and 40, where the RS (Register Select) pin of the LCD control determines if a command (RS = 0) or data (RS = 1) is being sent. Here we discuss the sendCmd function, which sends the command to the LCD. The command to be sent is placed on PORTB at line 28. The LCD control line RW (Read/Write) is set low at line 30 to indicate the write operation to LCD and then a delay at line 31 is called to allow time for the internal operation of the LCD. The command/data on the LCD data bus is transferred to the LCD when control line E rises from 0 to 1 at line 32 (which was first set low at line 67). After a short delay at line 33, E is set low for the next write operation at line 34.

3.8.2 PIC16F627A LCD Display – Assembler Program

The assembler program for the LCD display given in Listing 3.4 consists of six parts described below:

- Part 1 – lines 11 to 26 – assigns labels and memory locations to variables used in the program.
- Part 2 (sendData) – lines 33 to 42 – sends the character to be displayed to PORTA.
- Part 3 (sendCmd) – lines 45 to 54 – sends a command to the LCD.
- Part 4 (init_LCD) – lines 57 to 77 – sets up the LCD for the mode required.
- Part 5 (START) – lines 80 to 127 – the main program, which prints the authors' first names on the two lines of the LCD
- Part 6 (delay) – lines 131 to 142 – produces variable time delays.

The first part of the program is self-explanatory so we turn our attention to the main part of the program. Lines 81 and 82 are used to turn off the ADC pins RA0, RA1 and RA2, which are used as control lines for the LCD. PORTA and PORTB are configured as outputs in lines 84 and 85. The LCD control lines are all set low (0) in lines 86 to 88. At line 90 a call is made to the init_LCD subroutine to set up the LCD mode. The first character to be displayed – H – is loaded into the W register in line 92 and the routine to display the character sendData is called on line 93. This operation is repeated for all the characters to be displayed. Note that at line 105, the W register is loaded with the NewLine code and at line 106 a command is sent to the LCD. At line 127 the program awaits and does nothing, keeping the current character set displayed on the LCD.

```
 1   ;*************************************
 2   ;   RB        7    6    5    4    3    2    1    0 *
 3   ;  Data Bus   dp   G    F    E    D    C    B    A *
 4   ;-------------------------------------------------- *
 5   ;   RA        0    1    2                           *
 6   ;  Controls   RS   RW   E                           *
 7   ;************************************************/
 8   include "p16f627A.inc"
 9
10   ;   LCD Control Lines
11   RS    EQU                    0; LCD Register-Select
12   RW    EQU                    1; LCD Read/Write
13   E     EQU                    2; LCD Enable
14
15   ;   LCD Commands
16   NewLine                EQU 0xC0
17   ClrDisp                EQU 0x01
18   TwoLine10dots          EQU 0x38
19   ScrOnCurOffBlinkOn     EQU 0x0F
20   IncCurDntMovDis        EQU 0x06
21   CurHome                EQU 0x02
22
23   ;   global  variables
24   T1    EQU      0x26
25   T2    EQU      0x27
26   T3    EQU      0x28
27
28   ;*************************************
29          org     0x0000    ; RESET vector
30   RST  goto     START
31
32   ;*************************************
33   sendData                ; data is in W reg
34          movwf    PORTB    ; put char into RAM
35          bsf      PORTA,RS ; data mode
36          bcf      PORTA,RW ; write operation
37          movlw    0x01     ; set up delay
38          call     Delay    ; delay 1 ms
39          bsf      PORTA,E
40          call     Delay    ; delay 1 ms
41          bcf      PORTA,E
42          return
```

Listing 3.4   PIC16F627A LCD display – assembler program

```
43
44      ; * * * * * * * * * * * * * * * * * * * * * * * * * * * * * * *
45      sendCmd                          ; Cmd is in W reg
46          movwf       PORTB            ; put char into RAM
47          bcf         PORTA,RS         ; command mode
48          bcf         PORTA,RW         ; write operation
49          movlw       0x01             ; set up delay
50          call        Delay            ; delay 1 ms
51          bsf         PORTA,E
52          call        Delay            ; delay 1 ms
53          bcf         PORTA,E
54          return
55
56      ; * * * * * * * * * * * * * * * * * * * * * * * * * * * * * * *
57      init_LCD
58          movlw       D'40'            ; 40 ms delay
59          call        Delay
60
61          movlw       TwoLine10dots
62          call        sendCmd
63          movlw       1                ; 1 ms delay
64          call        Delay
65          movlw       ScrOnCurOffBlinkOn
66          call        sendCmd
67          movlw       1                ; 1 ms delay
68          call        Delay
69          movlw       ClrDisp
70          call        sendCmd
71          movlw       1                ; 1 ms delay
72          call        Delay
73          movlw       CurHome
74          call        sendCmd
75          movlw       1                ; 1 ms delay
76          call        Delay
77          return
78
79      ; * * * * * * * * * * * * * * * * * * * * * * * * * * * * * * *
80      START
81          movlw       0x07
82          movwf       CMCON            ; turn ADCs OFF
83          movlw       0x00             ; Set the Port pins
84          tris        PORTA            ; to outputs.
85          tris        PORTB
86          bcf         PORTA,RW         ; LCD_RW = 0
87          bcf         PORTA,RS         ; LCD_RS = 0
88          bcf         PORTA,E          ; LCD_E = 0
89
90          call        init_LCD         ; set up the LCD
91
92          movlw       'H'
93          call        sendData
```

Listing 3.4   (*continued*)

```
94        movlw    'a'
95        call     sendData
96        movlw    's'
97        call     sendData
98        movlw    's'
99        call     sendData
100       movlw    'a'
101       call     sendData
102       movlw    'n'
103       call     sendData
104
105       movlw    NewLine        ; go to beginning
106       call     sendCmd        ; of the next line
107
108       movlw    'B'
109       call     sendData
110       movlw    'r'
111       call     sendData
112       movlw    'a'
113       call     sendData
114       movlw    'n'
115       call     sendData
116       movlw 'i'
117       call     sendData
118       movlw    's'
119       call     sendData
120       movlw    'l'
121       call     sendData
122       movlw    'a'
123       call     sendData
124       movlw    'v'
125       call     sendData
126
127 AGAIN goto    AGAIN             ; wait for ever
128
129  * * * * * * * * * * * * * * * * * * * * * * * * * * * * * * * *
130  ;number of ms in W reg
131 Delay movwf   T1                ; 1 cycle
132 DelT1 movlw   D'2'              ; 1    "
133       movwf   T2                ; 1    "
134 DelT2 movlw   D'164'            ; 1    "
135       movwf   T3                ; 1    "
136 DelT3 decfsz  T3,1              ; 2    "
137  goto         DelT3             ; 2    "
138  decfsz       T2,1              ; 1    " repeat not 0
139  goto         DelT2             ; 2    "
140  decfsz       T1,1              ; 1    " repeat not 0
141  goto         DelT1             ; 2    "
142  return                         ; 2    "
```

Listing 3.4 (*continued*)

The init_LCD subroutine works by loading the W register with the command code, and then calling the sendCmd subroutine. The setup is self-explanatory but the reader should pay attention to the delays required after each command. The sendCmd at line 46 first puts the command code on the LCD data bus through PORTB and then at line 47 LCD control line RS is set low (to select the command register). At line 48 the LCD RW line is set low to indicate the write operation to the LCD. For a command to be written to the LCD, control line E should undergo a 0 to 1 transition and, as E is already 0, the transition to 1 is initiated at line 51. The E line is set low again at line 53, in order to be ready for the next write operation.

The sendData subroutine is exactly the same as sendCmd, except at line 35 (line 47 in sendCmd), where the register select line is set high to tell the LCD that the byte placed on the data bus is, in fact, data to be displayed and not the command to the LCD.

The Delay routine at line 131 has three nested loops. Their values are in variables T1, T2 and T3. T1 is the outermost loop variable and T3 is the innermost. First the value in the W register is transferred to variable T1 and the variables T2 and T3 are loaded with some values to give a delay of T1 ms. Comments give the number of cycles per instruction and each cycle (as we use internal clock of 4 MHz, and this is divided by 4) therefore takes 1 μS, and, as an exercise, it is left to the reader to calculate the time delay.

RS232 Interfaces

4.1 Introduction

In this chapter a circuit with an RS232 interface, based around the PIC16F627A microcontroller, is developed and its operation is described in detail. The necessary hardware is designed and discussed and programs using C and assembler languages are developed to make use of this circuitry. Programs described in this chapter can:

• send characters and strings to a PC with a serial port
• receive characters from the PC
• simultaneously send and receive characters to and from the PC.

The first programs in this chapter written using both C and assembler languages can send an array of characters to the PC, which uses Hyperterminal to display the received data. The next set of programs can receive the keys pressed on the keyboard of the PC and display the binary equivalent of the character received on eight LEDs connected to a PIC MCU. The final programs can send and receive simultaneously – when a key on the PC keyboard is pressed, information is sent to the PIC MCU and displayed on LEDs as before. The case of the received

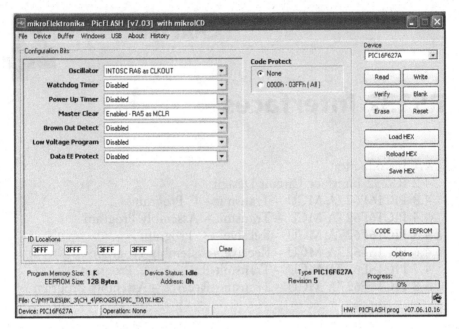

Figure 4.1 PIC16F627A programming setup

Source: MikroElectronika.

character is then changed (upper case to lower case or vice versa) and sent back to the PC to be displayed on Hyperterminal.

The PIC MCU in this chapter is programmed with settings shown in Figure 4.1, where pin RA5 is used as the Master Clear Pin and the rest of the settings are as before.

4.2 RS232 Interface Circuit Design

The circuit consists of several parts, which include the PIC MCU, the MAX232 RS232 driver/receiver, LED display and a simple RESET circuit. Figure 4.2 shows the interface developed using these components.

The MAX232 is a dual driver/receiver that includes a capacitor voltage generator to supply TIA/EIA-232F voltage levels from a single 5 V supply. Each receiver converts TIA/EIA-232F inputs to 5 V TTL/ CMOS levels. This application uses only one of the driver/receivers and the spare one could be used for further work when required. (In this diagram R2in, pin-8, is connected to ground to allow PCB design, but otherwise this connection is not required.)

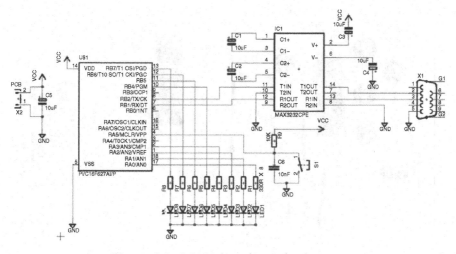

Figure 4.2 RS232 interface schematic

Four capacitors are used for voltage conversions. To connect to the PC a nine-pin D-type (female) connector was used, where pin 2 and pin 3 of the D-type are connected to the T1out (pin-14) and R1in (pin-13) of the MAX232 respectively. The PIC MCU interface to the MAX232 is achieved by connecting RB2/TX (pin 8) and RB1/RX (pin 7) of the PIC MCU to T1in (pin 11) and R1out (pin 12) respectively.

On the display side we have used only eight LEDs with current-limiting resistors. These are used to display the equivalent binary number of the character received by the board. As RB1 and RB2 are used by the MAX232, the lower nibble of the display is connected to RA0, RA1, RA2 and RA3 and the upper nibble is connected to pins RB4 to RB7.

An external RESET circuit, which is just a resistor, a capacitor with a push-button switch should be used for resetting the board at the beginning of the communication.

4.2.1 PIC12F629 Strip Board and PCB Designs

The strip board design shown in Figure 4.3 is used to build and implement this circuit. The PCB layout of the same circuit is shown in Figure 4.4. A single layer board with the components on top and the track side underneath was used in this project in order to produce a quick and simple implementation.

To see all the details of these diagrams, for readers who want to build this circuit using a strip board or PCB design, we recommend

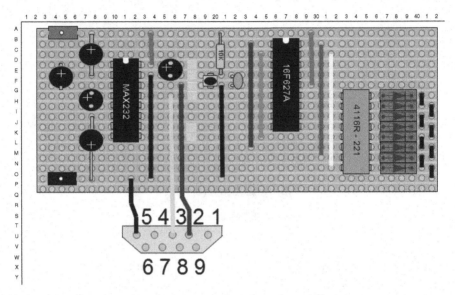

Figure 4.3 RS232 interface strip board design

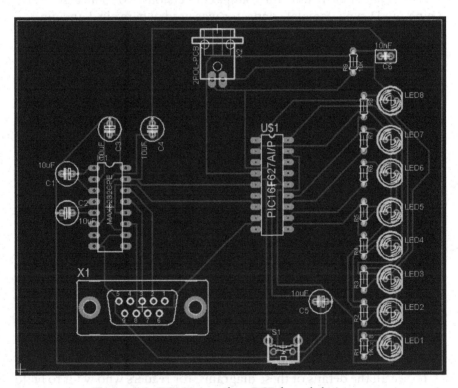

Figure 4.4 RS232 interface PCB board design

downloading the diagrams from the book's web site and displaying them on the computer screen or printing them out in a larger size.

4.2.2 PIC16F627A MCU – Universal Synchronous Asynchronous Receiver Transmitter (USART) Module

The USART is used for serial communication with peripheral devices with a serial port. The USART can be configured to operate in the following modes:

- Asynchronous (full duplex): communication in both directions. For example, a telephone is full duplex because parties can talk simultaneously. It is asynchronous, which means that the communication is not synchronized to a clock signal.
- Synchronous (half duplex): this refers to communication in one direction at a time. For example, a walkie-talkie is half duplex because only one party can talk at a time. It is synchronous, which means the communication is synchronized to a clock signal.

To set the USART for transmission, the following two PIC MCU registers need to be configured:

- TXSTA – transmit status and control register.
- RCSTA – receive status and control register.

Here we have a look at how to configure these two registers to enable us to asynchronously, (i) Transmit, (ii) Receive, and (iii) Transmit/Receive. Figure 4.5 shows the TXSTA and RCSTA registers.

In order to set up the USART to communicate with the Serial Port ENable pin (SPEN) of the RCSTA (bit 7) must be set. As this PIC MCU model uses RB2 and RB1 as Tx and Rx pins of the USART, RB2 and RB1 should also be configured as output and input respectively.

In addition to the above setup, to enable transmission through the USART one only needs to set bits BRGH (bit 2) and TXEN (bit 5) of the TXSTA register. In order to receive through the USART, bits BRGH (bit 2) of the TXSTA and CREN (bit 4) of the RCSTA register should be set.

4.3 PIC16F627A MCU – Transmit – C Program

Listing 4.1 shows the C program, which sets up the USART to transmit a character at a time and then continuously transmits the

TXSTA – TRANSMIT STATUS AND CONTROL REGISTER (ADDRESS: 98h)

R/W-0	R/W-0	R/W-0	R/W-0	U-0	R/W-0	R-1	R/W-0
CSRC	TX9	TXEN	SYNC	—	BRGH	TRMT	TX9D

bit 7   bit 0

bit 7	**CSRC**: Clock Source Select bit
bit 6	**TX9**: 9-bit Transmit Enable bit
bit 5	**TXEN**: Transmit Enable bit[1]
bit 4	**SYNC**: USART Mode Select bit
bit 3	**Unimplemented**: Read as '0'
bit 2	**BRGH**: High Baud Rate Select bit
bit 1	**TRMT**: Transmit Shift Register Status bit
bit 0	**TX9D**: 9th bit of transmit data. Can be parity bit.

RCSTA – RECEIVE STATUS AND CONTROL REGISTER (ADDRESS: 18h)

R/W-0	R/W-0	R/W-0	R/W-0	R/W-0	R-0	R-0	R-x
SPEN	RX9	SREN	CREN	ADEN	FERR	OERR	RX9D

bit 7   bit 0

bit 7	**SPEN**: Serial Port Enable bit
bit 6	**RX9**: 9-bit Receive Enable bit
bit 5	**SREN**: Single Receive Enable bit
bit 4	**CREN**: Continuous Receive Enable bit
bit 3	**ADEN**: Address Detect Enable bit
bit 2	**FERR**: Framing Error bit
bit 1	**OERR**: Overrun Error bit
bit 0	**RX9D**: 9th bit of received data (Can be parity bit)

Figure 4.5 PIC16F627A TXSTA and RCSTA registers

© Microchip Technology Inc. Reproduced with permission.

combination of characters from the PIC MCU to the PC. This program contains four functions. These are USART_setup() to set the USART for transmit mode only, putChar() to transmit one character, putString() to send a string of characters and finally main(), which calls the other functions to set up the USART and transmit characters and strings.

```
1   //************************************************
2   // Hardware        RB2 = Tx         RB1 = Rx          *
3   //------------------------------------------------*
4   // TXSTA TRANSMIT STATUS AND CONTROL REGISTER (98h) *
5   //   7    6      5      4     3     2      1      0   *
6   // CSRC  TX9  TXEN  SYNC    -   BRGH  TRMT  TX9D *
7   //************************************************
8
9   #include <pic16f62xa.h>
10
11  void USART_setup(void){
12   /* relates crystal freq to baud rate
13     BRGH=1, Fosc=4MHz
14     -----------------
15     Baud                            SPBRG
16     9600                    25
17     19200                   12
18     38400                    6
19     57600                    3
20     115200                   1
21     250000                   0
22   */
23     SYNC  = 0;                  // select asynchronous mode
24     SPBRG = 12;                 // baud rate 19200
25     BRGH  = 1;                  // high data rate for sending
26     SPEN  = 1;                  // enable serial port pins
27     TXIE  = 0;                  // disable tx interrupts
28     TX9   = 0;                  // 8-bit transmission
29     TXEN  = 1;                  // enable the transmitter
30  }
31
32  // writes a character to the serial port
33  void putChar(unsigned char ch){
34     while(!TXIF)               //set when TXREG is empty
35           ;
36     TXREG=ch;
37  }
38
39  // writes a string to the serial port
40  void putString(register const char *Str){
41     while((*Str)!=0) {
42       putChar(*Str);
43     if (*Str==13) putChar(10);
44     if (*Str==10) putChar(13);
45       Str++;
46   }
47  }
48
49  void main(void){
50     TRISB =0b00000010;
51
```

Listing 4.1   PIC16F627A transmit – C program

```
52    USART_setup();
53    while(1){
54       putChar('H');
55       putString("Hassan \n");
56       putString("Branislav \n");
57    }
58  }
```

Listing 4.1 (*continued*)

The first few lines of the program are comments to remind the programmer and other people intending to use this program about the relevant bits of the TXSTA register. The main part of the program starts at line 49. On line 50, PORTB is configured to have RB1 (Rx) as input and RB2 (Tx) as output. Then, at line 52, the USART_setup() function is called. The operation of this function is explained later in this section. An infinite, while(1) loop is defined between lines 53 to 57. The purpose of this loop is to demonstrate how to send a character or a combination of characters to the PC using putChar() and putString() functions.

The USART_setup() function starts at line 11; the function does not take or return any parameters. Commented lines 15 to 21 show the values that need to be put in the SPBRG register in order to achieve various baud rates. Control bits and the SPBRG register have been configured between lines 23 and 29 in the following way:

- SYNC bit is set to 0, to achieve operation in asynchronous mode
- SPBRG is set to 12, for 19200 baud rate
- BRGH is set to 1, to use high data rate
- SPEN is set to 1, to enable the serial port
- TXIE is set to 0, to stop transmit interrupt
- TX9 is set to 0, to use an 8-bit data mode, and
- TXEN is set to 1, to enable the transmit mode.

The putChar() function is defined between lines 33 and 37. This function receives a character (ch) from the main program and transmits this character through the USART. First, at line 34, a while() loop is used to test and make sure that TXREX, transmit register, is empty, by checking the flag bit TXIF. This bit remains at value 0 as long as the TXREX register is not empty. When the flag is set, the condition in the while loop (!TXIF) is no longer true and the program proceeds to line 36 to put character ch into TXREG, to be transmitted.

To send a whole string – a combination of characters – function putString() is used in our program. This function, defined between lines 40 to 47, receives a pointer to a string (*Str). This function uses C pointers to point to the characters in the string to be transmitted. First, at line 41, string contents are checked to make sure it is not the end of the string and, while this is the case, lines 42 to 45 are executed. At line 42, putChar() is called to transmit the character pointed to by the *Str pointer. Then a check is made for the end of the line and carriage return characters at lines 43 and 44. If they exist in the string, characters are sent. The pointer to the string is incremented to point to the next character in the string at line 45.

4.3.1 Transmit Program Communication with the PC

Testing of the PIC MCU transmit program can be done in two ways – using the Windows Hyperterminal application or the PIC_Terminal program. This program is available for download from the book's web site. In the case of Hyperterminal, some setting up is needed (explained later in this section) but in the case of PIC_Terminal, no special configuration is required.

4.3.1.1 Communication with the PC using Hyperterminal

On most installations of Windows, the Hyperterminal program can be found by selecting Programs – Accessories – Communications from the Start menu. When the Hyperterminal program is run, a window such as that shown in Figure 4.6 is displayed.

Figure 4.6 Hyperterminal program – Connection Description window
Microsoft product screen shot reprinted with permission from Microsoft Corporation.

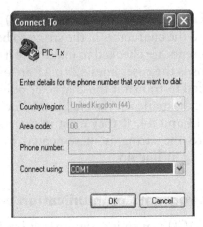

Figure 4.7 Hyperterminal program – Connect To window

Microsoft product screen shot reprinted with permission from Microsoft Corporation.

To set up the program, type in a name for the session (PIC_Tx) and click the OK button, as shown in Figure 4.6. A new window appears, shown in Figure 4.7.

To progress further through the setup, click the OK button on the Hyperterminal 'Connect To' window, if your RS232 is connected to the COM1 port, otherwise choose an appropriate port and click OK to fill in the COM1 (or other selected com port) properties window shown in Figure 4.8.

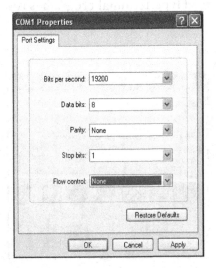

Figure 4.8 Hyperterminal program – COM1 properties window

Microsoft product screen shot reprinted with permission from Microsoft Corporation.

In the dialogue box, bits per second should be set to 19 200, and Flow Control to None. The rest of the settings should be correct, as shown in Figure 4.8. Press Apply and then OK. If the PIC MCU transmit program is running, the strings sent to the PC by the PIC_Tx program should be displayed in the Hyperterminal window as shown in Figure 4.9. If the string is not displayed, pressing the reset button on the PIC board several times should rectify the problem.

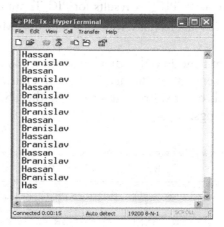

Figure 4.9 PIC_Tx results

Microsoft product screen shot reprinted with permission from Microsoft Corporation.

4.3.1.2 Communication with the PC using the PIC_Terminal Program

To use the PIC_Terminal program, the reset button on the PIC board should be held down, the PIC_Terminal activated and then the reset button released. A window like the one shown in Figure 4.10 should be displayed. In case of any errors during the initial run of the PIC_Terminal program, the message window should be closed and the whole operation repeated.

4.4 PIC16F627A MCU – Transmit – Assembly Program

Listing 4.2 shows the assembly program that sets up the USART to transmit a character at a time and continuously transmits the names of the authors to the PC. This program contains three subroutines and the Start section of the code. These are USART_Setup, used to set the

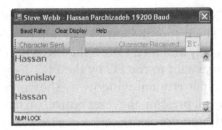

Figure 4.10 PIC_Tx results for PIC_Terminal.

Microsoft product screen shot reprinted with permission from Microsoft Corporation.

USART for transmit mode only, putChar to transmit one character, putString to send a predefined string of characters, and finally START, which calls other subroutines to set up the USART and transmit characters and strings.

```
 1  ;****************************************************
 2  ;Hardware          RB2 = Tx          RB1 = Rx        *
 3  ;--------------------------------------------------*
 4  ;TXSTA TRANSMIT STATUS AND CONTROL REGISTER (98h) *
 5  ; 7    6     5     4     3     2     1     0    *
 6  ;CSRC TX9  TXEN  SYNC   -    BRGH  TRMT  TX9D *
 7  ;****************************************************
 8  ;include "p16f627A.inc"
 9
10  ; global variables
11  charTx  equ  0x20
12
13  ;****************************************************
14  org        0x0000         ; RESET vector
15  RST        goto START
16
17  ;****************************************************
18  USART_Setup
19  ; relates crystal freq to baud rate
20  ;
21  ; BRGH=1,        Fosc=4MHz
22  ; ------------------
23  ; Baud          SPBRG
24  ; 9600          25
25  ; 19200         12
26  ; 38400         6
27  ; 57600         3
28  ; 115200        1
29  ;250000         0
30  ;
```

Listing 4.2 PIC16F627A transmit – assembly program

```
31    bsf       STATUS,RP0          ;select page 1
32    bcf       TXSTA, SYNC         ;select asynchronous mode
33    movlw     0x0C                ;0x0C = 12
34    movwf     SPBRG               ;baud rate 19200
35    bsf       TXSTA, BRGH         ;data rate for sending
36    bcf       PIE1, TXIE          ;disable tx interrupts
37    bcf       TXSTA, TX9          ;8-bit transmission
38    bsf       TXSTA, TXEN         ;enable the transmitter
39    bcf       STATUS,RP0          ;select page 0
40    bsf       RCSTA, SPEN         ;enable serial port pins
41    return
42    ;***************************************************
43    ;send a character to the serial port
44    putChar
45    bcf       STATUS,RP0          ;select page 0
46 CHK btfss    PIR1, TXIF          ;check if ready to transmit
47    goto      CHK
48    movf      charTx,0            ;W = charTx
49    movwf     TXREG               ;Transmit
50    return
51
52    ;***************************************************
53    ;send the string to the serial port
54    putString
55    movlw     'a'
56    movwf     charTx
57    call      putChar
58    movlw     's'
59    movwf     charTx
60    call      putChar
61    movlw     's'
62    movwf     charTx
63    call      putChar
64    movlw     'a'
65    movwf     charTx
66    movwf     charTx
67    call      putChar
68    movlw     'n'
69    movwf     charTx
70    call      putChar
71    movlw     0x0A
72    movwf     charTx
73    call      putChar
74    movlw     0x0D
75    movwf     charTx
76    call      putChar
77    return
78
79    ;***************************************************
80    START
81    bsf       STATUS,RP0          ;select page 1
82    movlw     0x02
83    movwf     TRISB               ;RB1 input, the rest outputs
```

Listing 4.2 (*continued*)

```
84      call            USART_Setup
85      bcf             STATUS,RP0          ;select page 0
86
87 AGAIN
88      movlw           'H'
89      movwf           charTx
90      call            putChar             ;transmit H
91      call            putString           ;transmit assan
92      goto            AGAIN               ;repeat again
93 end
```

Listing 4.2 (*continued*)

The first few lines are comments explaining the individual bits of TXSTA register. The main part of the program starts at line 81, where the register bank is switched to page 1 to enable access to TRISB register. The working register W is loaded with hexadecimal 2 (binary 0000 0010) at line 82, and at line 83, PORTB is configured to have RB1(Rx) as input and RB2 (Tx) as output. The USART_Setup subroutine is called at line 84. This subroutine is discussed in more detail later in this section. Line 85 switches back register banks to page 0. Lines 88 to 91 are executed continuously to send authors' names to the PC, using the putChar and putString subroutines.

The USART_Setup subroutine starts at line 18, but lines 19 to 30 are comments to show the values to be put in the SPBRG register in order to achieve various baud rates. At lines 31 to 40 the control bits and SPBRG register have been configured. The setting is the following, which is similar to the C version of this program:

- SYNC bit is set to 0, to operate in asynchronous mode
- SPBRG is set to 12, for 19 200 baud rate
- BRGH is set to 1, to use a high data rate
- SPEN is set to 1, to enable the serial port
- TXIE is set to 0, to stop transmit interrupt
- TX9 is set to 0, to use 8-bit data mode
- TXEN is set to 1, to enable the transmit mode.

At lines 43 to 50 the putChar subroutine transmits the charTx through USART. First at line 45, page 0 of the register banks is selected and then lines 46 to 48 are used to check and make sure that the TXREX transmit register is empty, by checking the flag bit TXIF, which is 0 when TXREX is empty. When the flag is set, the program proceeds to line 48 to move charTx into the W and then into TXREX, for transmission.

The putString subroutine at lines 54 to 77 transmits the combination of characters, by moving the characters to be sent into the charTx variable and then calling the putChar subroutine for transmission.

To test this program, the reader should follow the procedure explained and discussed in the previous section for the case of the C version of the same program.

4.5 PIC16F627A MCU – Receive – C Program

Listing 4.3 shows the C program which sets up the USART to receive a single character at a time and continuously display the pattern corresponding to the received character on the PIC MCU board LEDs. This program contains three functions including main(). These are USART_setup(), which sets the USART for receive mode only, getChar(), which returns the received character and finally main() , which calls other functions to set up the USART, and display the characters received.

```
 1  // ************************************************
 2  // Hardware RB2 = Tx RB1 = Rx                      *
 3  // ------------------------------------------------*
 4  //RCSTA RECEIVE STATUS AND CONTROL REGISTER (18h)  *
 5  //   7     6     5     4     3     2     1     0    *
 6  // SPEN  RX9   SREN  CREN  ADEN  FERR  OERR  RX9D   *
 7  // ************************************************
 8
 9
10  #include <pic16f62xa.h>
11
12  void USART_setup(void){
13  /* relates crystal freq to baud rate
14
15     BRGH=1,    Fosc=4MHz
16     ------------------
17     Baud                SPBRG
18     9600                25
19     19200               12
20     38400               6
21     57600               3
22     115200              1
23     250000              0
24  */
25     SYNC = 0;               // select asynchronous mode
26     SPBRG = 12;                // baud rate 19200
```

Listing 4.3   PIC16F627A receive – C program

```
27  BRGH = 1;           // data rate for sending
28  SPEN = 1;           // enable serial port pins
29  RCIE = 0;           // disable Rx interrupts
30  RX9  = 0;           // 8-bit reception
31  CREN = 1;           // Enables continuous receive
32 }
33
34 //gets a character from the serial port without timeout
35 unsigned char getChar (void) {
36    while(!RCIF)
37        ;
38    return RCREG;
39 }
40
41  void main (void) {
42  unsigned char charRx;
43
44  TRISB = 0b00000010; // RB1 set as input, the rest outputs
45  CMCON = 0x0F;       // RA0, 1, 2 and 3 as digital I/Os
46  TRISA = 0xF0;       // set to outputs
47
48
49  USART_setup();
50  while(1) {
51   charRx = getChar();
52   PORTA = charRx & 0x0F; // display lower 4 bits on LEDs
53   PORTB = charRx & 0xF0; // display higher 4 bits on LEDs
54   }
55 }
```

Listing 4.3   (*continued*)

Here, as in the case of the transmit program, comments detailing the bits of the RCSTA register are provided in the first seven lines of the program. The main part starts at line 44, where PORTB is configured to have RB1(Rx) as input and RB2 (Tx) as output. For the receive program, RA0, RA1, RA2 and RA3 are also used as outputs to drive the LEDs on the PIC board, so they should be configured as well, but these pins can also be configured as analog inputs (multiplexed) so the CoMparator CONtrol register (CMCON) comparators are turned off. RA0–RA3 are configured as digital type outputs in lines 45 and 46. Line 49 calls USART_setup(), which is explained later in this section. Between lines 50 to 54 a while(1) loop is executed, which continuously reads the character received and displays it through two nibbles of PORTB (RB7–RB4) and PORTA (RA3–RA0) on the PIC board LEDs.

The USART_setup() starts at line 12, and is very similar to the corresponding function explained in the previous section where the 'transmit' program was discussed. The main difference is that control bits RCIE, RX9 and CREN are configured in lines 29, 30 and 31 instead of bits TXIE, TX9 and TXEN. Here they are configured as:

RCIE is set to 0, to stop receive interrupt
RX9 is set to 0, to use 8-bit data mode
CREN is set to 1, to enable the receive mode.

The getChar() function defined between lines 35 to 39 has no input parameters but returns a single value – the character received from the PC. At line 36 of this function, a while() loop is used to check the value of the flag bit RCIF. If this value is 0 no character is yet received. When the flag is set, the while loop condition is no longer true and the program proceeds to line 38 to return the character received in the RCREG register.

4.5.1 Receive Program Communication with the PC

To test the receive program, Hyperterminal or PIC_Terminal tools can be used. The Hyperterminal program should be configured as already explained in section 4.3.1.1. Typing characters (such as A, B, C . . .) should cause the display of the corresponding binary pattern on the LEDs attached to PIC. If, for example, A is pressed, the LEDs would show the 0100 0001 pattern (41 hex; the ASCII code for A). The same result should be achieved when using the PIC_Terminal program during the test.

4.6 PIC16F627A MCU – Receive – Assembly Program

Listing 4.4 shows the assembly program which sets up USART to receive one character at a time and continuously display the character on the PIC board LEDs. This program contains two subroutines – USART_setup, which configures the USART for receive mode only, and getChar, which gets the character received and places it in the variable charRx.

```
1 ; ****************************************************
2 ; Hardware    RB2 = Tx                RB1 = Rx          *
3 ; --------------------------------------------------- *
4 ; RCSTA RECEIVE STATUS AND CONTROL REGISTER (18h) *
5 ;    7      6      5      4      3      2      1      0     *
6 ; SPEN   RX9   SREN   CREN   ADEN   FERR   OERR   RX9D *
7 ; ****************************************************
```

Listing 4.4   PIC16F627A receive – assembly program

```
 8    include "p16f627A.inc"
 9
10 ; global variables
11 charRx equ 0x20
12 ; **********************************************
13 org        0x0000         ; RESET vector
14 RST        goto   START
15
16 ; **********************************************
17 USART_Setup
18 ; relates crystal freq to baud rate
19 ;
20 ; BRGH=1, Fosc=4MHz
21 ; -----------------
22 ; Baud              SPBRG
23 ; 9600              25
24 ; 19200             12
25 ; 38400             6
26 ; 57600             3
27 ; 115200            1
28 ; 250000            0
29 ;
30     bsf    STATUS,RP0     ; select page 1
31     bcf    TXSTA, SYNC    ; select asynchronous mode
32     movlw  D'12
33     movwf  SPBRG          ; baud rate 19200
34     bsf    TXSTA, BRGH    ; data rate for sending
35     bcf    STATUS, RP0    ; select page 0
36     bsf    RCSTA, SPEN    ; enable serial port pins
37     bcf    RCSTA, RX9     ; 8-bit reception
38     bsf    RCSTA, CREN    ; Enables continuous receive
39     bsf    STATUS,RP0     ; select page 1
40     bcf    PIE1, RCIE     ; disable rx interrupts
41     return
42 ; **********************************************
43 ; gets a character from the serial port without timeout
44 getChar
45         bcf    STATUS,RP0     ; select page 0 for PIR1
46 chk     btfss  PIR1, RCIF     ; wait till data received
47         goto   chk
48         movf   RCREG,0        ; read the data received
49         movwf  charRx         ; save it
50         return
51 ; **********************************************
52
53     START
54         bsf    STATUS, RP0    ; select page 1
55         movlw  0x02
56         movwf  TRISB          ; RB1 is input, the
                                   rest outputs
57         bcf    STATUS,RP0     ; select page 0
58         movlw  0x0F
```

Listing 4.4   (*continued*)

```
59          movwf   CMCON           ; RA0,1,2 and 3 as
                                      digital I/Os
60          bsf     STATUS,RP0      ; select page 1
61          movlw   0xF0
62          movwf   TRISA           ; set to outputs
63          call    USART_Setup
64          bcf     STATUS,RP0      ; select page 0
65
66  AGAIN
67          call    getChar         ; read serial port
68          movf    charRx,w
69
70          movf    charRx,w        ;
71          andlw   0x0F            ; PORTA = charRx & 0x0F
72          movwf   PORTA           ;
73
74          movf    charRx,w        ;
75          andlw   0xF0            ;PORTB = charRx & 0xF0
76          movwf   PORTB           ;
77
78          goto    AGAIN           ;   repeat   again
79          end
```

Listing 4.4 (*continued*)

The program starts at line 54, where PORTB is configured to have RB1(Rx) as input and RB2 (Tx) as output. The Receive program uses RA0, RA1, RA2 and RA3 pins as output pins to drive the LEDs on the PIC board and they should be configured as well. In some other applications these pins can also be used as analog inputs (multiplexed). To prevent this, CoMparator CONtrol register (CMCON) is turned off and RA0–RA3 are configured as digital I/Os at lines 57 to 59. Lines 60, 61 and 62 configure those bits to be digital outputs. Line 63 calls the USART_Setup subroutine, discussed later in this section. Between lines 66 to 78 an endless loop is executed, which continuously calls the getChar subroutine and displays the received character using the upper nibble of PORTB (RB7–RB4) and lower nibble of PORTA (RA3–RA0) to light up LEDs on the PIC board.

The USART_Setup subroutine starts at line 17 and is very similar to the subroutine used previously for the transmit program. The only difference is the configuration of bits RCIE, RX9 and CREN instead of TXIE, TX9 and TXEN. Here those bits are configured as:

- RCIE is set to 0, to stop receive interrupt
- RX9 is set to 0, to use 8-bit data mode
- CREN is set to 1, to enable the receive mode.

At lines 44 to 50, the `getChar` subroutine places the character received from the PC in the variable `charRx`. First line at, 45, page 0 is selected to have access to the PIR1 register. Then a check is continuously made at line 46, to ensure that a character is received, by checking the flag bit RCIF, which is 0, when no character is received. When the flag is set, the next instruction at line 47 is skipped and the program proceeds to line 48 to place the character received in RCREG into variable `charRx`.

To test this program, the reader should follow the procedure already explained for the C version of the same program.

4.7 PIC16F627A MCU – Transmit-Receive – C Program

Listing 4.5 shows the C program, which continuously receives a character from the PC in a lower (or upper) case, displays the character on the PIC MCU board LEDs, changes the case of the character and transmits it back to the PC.

```
 1   // ************************************************
 2   // Hardware       RB2 = Tx          RB1 = Rx           *
 3   // ---------------------------------------------- *
 4   // TXSTA TRANSMIT STATUS AND CONTROL REGISTER (98h) *
 5   //   7      6      5     4    3     2      1      0 *
 6   // CSRC    TX9    TXEN   SYN      BRGH   TRMT   TX9D *
 7   // ************************************************
 8   // RCSTA RECEIVE STATUS AND CONTROL REGISTER (18h) *
 9   //   7      6      5     4    3     2      1      0 *
10   // SPEN   RX9    SREN  CREN   ADEN  FERR   OERR   RX9D *
11   // ************************************************
12   #include <pic16f62xa.h>
13
14   void USART_setup(void){
15   /* relates crystal freq to baud rate
16
17   BRGH=1,    Fosc=4MHz
18   -------------------
19   Baud          SPBRG
20   9600          25
21   19200         12
22   38400         6
23   57600         3
24   115200        1
25   250000        0
26   */
27     SYNC = 0;        // select asynchronous mode
28     SPBRG = 12;      // baud rate 19200
29     BRGH = 1;        // high data rate for sending
```

Listing 4.5 PIC16F627A transmit–receive – C program

```
30    SPEN = 1;          // enable serial port pins
31
32    TXIE = 0;          // disable tx interrupts
33    TX9 = 0;           // 8-bit transmission
34    TXEN = 1;          // enable the transmitter
35
36    RCIE = 0;          // disable rx interrupts
37    RX9 = 0;           // 8-bit reception
38    CREN = 1;          // Enables continuous receive
39 }
40
41 // writes a character to the serial port
42 void putChar(unsigned char ch){
43 while(!TXIF)    //set when TXREG is empty
44          ;
45 TXREG=ch;
46 }
47
48 //gets a character from the serial port without timeout
49 unsigned char getChar(void){
50     while(!RCIF)
51          ;
52     return RCREG;
53 }
54
55 void main(void){
56
57     unsigned char charRxTx;
58
59     TRISB =0b00000010; // RB1 set as input, the rest
                             outputs
60     CMCON = 0x0F;        // RA0,1,2 and 3 as I/Os
61     TRISA = 0xF0;        // set to outputs
62
63     USART_setup();
64     while(1){
65          charRxTx = getChar();
66          PORTA = charRxTx & 0x0F;
67          PORTB = charRxTx & 0xF0;
68          charRxTx = charRxTx ^ 0x20;
69          putChar(charRxTx);
70     }
71 }
```

Listing 4.5 *(continued)*

Important parts of this program, shown in Listing 4.5, have already been discussed in the previous sections where transmit and receive programs have been developed. Line 68, where the character case is changed, deserves additional attention. This line implements the 'exclusive OR' ('ex-or') operation between the variable charRxTx and hexadecimal value of 0x20. The difference between the ASCII codes for lower and upper case letters is 32 (hexadecimal 0x20). For example, the ASCII code for letter A is 0x41 and for letter a it is 0x61. Looking at the binary values of those hexadecimal numbers and implementing the 'ex-or' operation makes the reason for using 'ex-or' clearer:

A = 41 hex = 0100 0001 binary   a = 61 hex = 0110 0001 binary
Ex-or with 0010 0000   Ex-or with 0010 0000

            0110 0001 => a   0100 0001 => A

As can be seen above, using 'ex-or' with any letter with hex 20 would change the case of the letter.

4.7.1 Transmit-Receive Program Communication with the PC

We can again use any of two test tools, Hyperterminal or PIC_Terminal, to examine the performance of this program. Running those applications and typing characters like A, B and C should light up the LEDs on the board with the binary code for the characters sent and at the same time the case of the sent character would be changed and transmitted to the PC to be displayed in Hyperterminal or the PIC_Terminal window.

4.8 PIC16F627A MCU – Transmit-Receive – Assembly Program

Listing 4.6 shows the assembly program, which continuously receives a character from the PC in a lower (or upper) case, displays the character on the PIC board LEDs and transmits the character received in upper (or lower) case back to the PC.

```
1  ; **************************************************
2  ; Hardware      RB2 = Tx              RB1 = Rx         *
3  ; --------------------------------------------------- *
4  ; TXSTA TRANSMIT STATUS AND CONTROL REGISTER (98h) *
5  ;    7      6     5      4      3     2      1      0  *
6  ; CSRC   TX9   TXEN   SYNC    _    BRGH   TRMT   TX9D  *
7  ; --------------------------------------------------- *
8  ; RCSTA  RECEIVE  STATUS  AND  CONTROL  REGISTER (18h) *
9  ;    7      6     5      4      3     2      1      0  *
```

Listing 4.6   PIC16F627A transmit-receive – assembly program

```
10 ; SPEN  RX9  SREN  CREN  ADEN  FERR  OERR  RX9D *
11 ; *******************************************
12   include "p16f627A.inc"
13
14 ; global variables
15 charRxTx equ 0x20
16
17 ; *******************************************
18          org       0x0000      ; RESET vector
19 RST      goto      START
20
21 ; *******************************************
22 USART_Setup
23 ; relates crystal freq to baud rate
24 ;
25 ; BRGH=1, Fosc=4MHz
26 ; -----------------
27 ; Baud     SPBRG
28 ; 9600       25
29 ; 19200      12
30 ; 38400       6
31 ; 57600       3
32 ; 115200      1
33 ;250000       0
34 ;
35    bsf     STATUS,RP0     ; select page 1
36    bcf     TXSTA, SYNC    ; select asynchronous mode
37    movlw   0x0C
38    movwf   SPBRG          ; baud rate 19200
39    bsf     TXSTA, BRGH    ; data rate for sending
40    bcf     PIE1, TXIE     ; disable tx interrupts
41    bcf     TXSTA, TX9     ; 8-bit transmission
42    bsf     TXSTA, TXEN    ; enable the transmitter
43    bcf     PIE1, RCIE     ; disable rx interrupts
44    bcf     STATUS, RP0    ; select page 0
45    bcf     RCSTA, RX9     ; 8-bit reception
46    bsf     RCSTA, CREN    ; Enables continuous receive
47    bsf     RCSTA, SPEN    ; enable serial port pins
48        return
49
50 ; *******************************************
51 ; gets a character from the serial port without timeout
52   getChar
53       bcf      STATUS, RP0   ; select page 0 for PIR1
54 chk      btfss PIR1, RCIF    ; wait till data received
55       goto chk
56       movf RCREG,0           ; read the data received
57       movwf charRxTx         ; save it
58       return
59
60
61
```

Listing 4.6 (*continued*)

```
62  ; ***********************************************
63  ; send a character to the serial port
64  putChar
65         bcf      STATUS,RP0    ; select page 0
66  CHK    btfss PIR1, TXIF   ; check if ready to transmit
67         goto     CHK
68         movf     charRxTx,0    ; W = charRxTx
69         movwf    TXREG         ; Transmit
70         return
71
72  ; ***********************************************
73  START
74         bsf      STATUS,RP0    ; select page 1
75         movlw    0x02
76         movwf    TRISB         ; RB1 input, the rest outputs
77         bcf      STATUS,RP0    ; select page 0
78         movlw    0x0F
79         movwf    CMCON         ; RA0,1,2 and 3 as I/Os
80         bsf      STATUS,RP0    ; select page 1
81         movlw    0xF0
82         movwf    TRISA         ; set to outputs
83         call     USART_Setup
84         bcf      STATUS,RP0    ; select page 0
85  AGAIN
86         call     getChar       ; read serial port
87         movf     charRxTx,w    ;
88         andlw    0x0F          ; PORTA = charRx & 0x0F
89         movwf    PORTA         ;
90         movf     charRxTx,w    ;
91         andlw    0xF0          ; PORTB = charRx & 0xF0
92         movwf    PORTB         ;
93         movlw    0x20
94         xorwf    charRxTx,f
95         call     putChar       ; transmit data
96         goto     AGAIN         ; repeat again
97         end
```

Listing 4.6 (*continued*)

As in the previous section, the program presented here is a combination of transmit and receive programs already discussed in sections 4.4 and 4.6. Moreover, the case of the character is changed before its transmission to PC. This is done on line 94 by 'ex-or'-ing the corresponding ASCII value of the character with hexadecimal value 0x20 and thus toggling bit 6 of the ASCII code. This procedure has also been explained in the previous section.

To test this program the reader should follow the steps already explained for the C version of the same program.

Interfacing PICs with the Analog World

5.1 Introduction

So far, in this book, we have been developing and discussing the interfaces between the microcontroller and the digital world. All of the inputs and outputs to and from the PIC were in digital format – patterns consisting of groups of eight or more bits. The real world, unfortunately (from the PIC point of view), is not usually digital. It is analog. Analog quantities are continuous in nature whereas digital ones are discrete. Continuously varying voltage at the output of the microphone or some other transducer is a good example of an analog signal. Analog signals cannot be input directly to the microcontroller and processed; they need to be converted into digital form. An analog electrical voltage signal can be digitized using an electronic circuit called an analog-to-digital converter or, in short form, ADC. This circuit will generate a digital output as a stream of binary numbers whose values represent the electrical voltage input to the device at each sampling instant. In this chapter we will not delve into details of ADC operation but will simply consider it as a device that takes the analog signal at its input and produces a digitized version of this signal at its output. This simplified approach to conversion of analog signal to digital signal is shown in Figure 5.1.

Some of the PIC microcontrollers contain an ADC, so very little additional hardware needs to be added to the PIC-based system in order

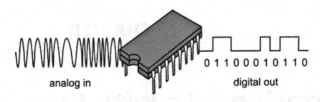

analog in   digital out

Figure 5.1 Analog to digital converter and its operation

to read and process analog-type signals. In this chapter we will explain how to configure and use this feature using the PIC18F252 device. The PIC18F252 pin diagram is shown in Figure 5.2. Pins RA0, RA1, RA2, RA3 and RA5 can be used as digital input/output pins as we have already shown in previous chapters and programs in this book but the PIC can also be configured to use those same pins to sense and process analog voltages. In this case those pins are referred to as AN0, AN1, AN2, AN3 and AN4 pins. On a power-up reset those pins default to analog inputs.

In this chapter we will demonstrate a few additional features of the PIC MCUs, using an advanced member of the PIC family – the PIC18F252. First, we will write a program that converts the analog signal supplied to one of the PIC pins to a digital version and checks the value of this signal against some preset voltage value. This basic program will then be extended to include some timing capabilities – the value of analog voltage will be sampled and converted to a digital version at regular time intervals, thus achieving a constant sampling and an AD conversion rate. An important additional feature of the PIC MCU is the internal timer and this will be used and configured to accurately measure

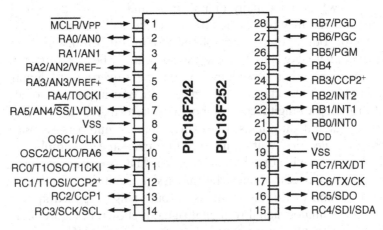

Figure 5.2 Pin diagram for PIC18F242 and PIC18F252 models (28-pin package)

© Microchip Technology Inc. Reproduced with permission.

those intervals called sampling periods. The program to achieve this task is the second program in this chapter. This program is not particularly efficient as it forces the PIC MCU to spend most of the time monitoring the state of the timer and initiating conversion of analog voltage at a regular time period according to the state of this timer. This basic timer program can be improved by using the interrupt feature of the PIC MCU. The interrupt feature enables the timer to operate in the background and interrupts the program operation at particular time intervals during normal program execution. This configuration therefore enables the CPU to work on some other task, thus significantly improving the efficiency of the program. The third program presented in this chapter is the program to configure and enable the timer interrupt operation on the PIC18F252. It is recommended that the reader spends some time understanding and testing the operation of this program. Many other interrupt sources exist on the PIC18F252 as well as most of the other PIC family members. Those sources can be used in other more advanced PIC projects, not discussed in this book. By understanding how timer interrupt and interrupt features in general operate, the reader will be able to consult other PIC books and understand the operation of more advanced, interrupt-driven programs written for PICs or even other similar microcontrollers. In this chapter the PIC18F252 was programmed with the options shown in Figure 5.3.

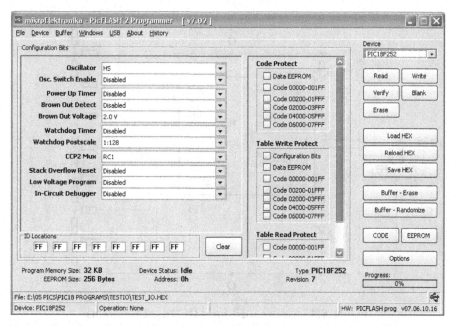

Figure 5.3 Options used for the programming of the PIC18F252 in this chapter

Source: MikroElectronika.

5.2 Hardware Description

The schematic diagram for the circuit used to test the programs developed in this chapter is shown in Figure 5.4. We leave it to the reader to decide on the way to implement this circuit in hardware in order to test programs in this chapter. The whole system is based around a PIC18F252 microcontroller, which is clocked with a 16 MHz crystal connected to pins 9 and 10. The analog signal is sensed via Pin 2 (AN0).

The first three programs basically perform the same task – they measure the voltage applied to pin AN0 and switch on the LED connected to RC3 if the measured voltage level exceeds the threshold level set in the program. If the measured voltage level falls bellow this threshold value the LED is switched off. An additional feature of timing – sampling the input voltage at the output – is added to the second program from this group and improved timing using the built-in timer of PIC18F252 is demonstrated in third program.

Finally, the fourth program in this chapter simply reads the voltage from AN0 and outputs the corresponding 8-bit digital value to PORTB. An R–2R ladder network of resistors connected to PORTB (R0–R7) then converts this digital value back into an analog form. The

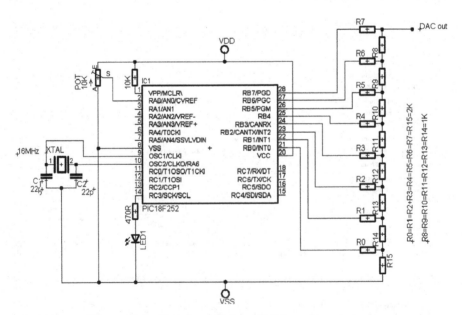

Figure 5.4 Circuit used for ADC-based programs

R–2R network is implemented by a set of resistors with two values; with one of them double the other. We have used $1\,k\Omega$ and $2\,k\Omega$ resistor values, although any other combination would produce similar results. This network therefore represents a simple DAC (digital-to-analog circuit) and is used to develop and test a simple 'talkthrough'-type program where the input signal just goes through the microcontroller, being converted to a digital and back to an analog version in the process. We will demonstrate how to use this program to perform some further processing of digitized input signal (digital signal processing) before this signal is converted back to the analog form, in the final chapter of the book.

The DAC circuit used in this project is an 8-bit one, but more bits can be easily added, to provide a higher resolution output signal, by providing more R–2R branches to the circuit in Figure 5.4. As the built-in ADC on the PIC18F252 is a 10-bit device, using the 8-bit DAC actually reduces the accuracy of our system. A good exercise for the reader would therefore be to design and build a 10-bit R–2R-based DAC and modify the 'talkthrough' program given at the end of this chapter to suit this improved DAC. It is also possible to obtain a ready-made DAC circuit as a separate chip and interface it to output pins of the PIC microcontroller in order to obtain a better quality analog signal.

5.3 Level Indicator Program and Advanced Simulator Features

The first program we developed to demonstrate the use of the ADC has the task of measuring the voltage supplied via pin AN0. If the measured voltage exceeds a certain preset value, for example $1\,V$, an LED connected to RC3 of PORTC will be lit – pin RC3 will be set to logic 1. If the level of measured voltage signal stays below $1\,V$ or falls below that value during the program operation, pin RC3 will be reset back to logic 0 and the 'alarm' LED attached to that pin will be switched off.

Before discussing this program further, it is useful to understand how to configure the ADC module to use pin RA0 as an analog input. Registers ADCON0 and ADCON1 are used to set up and configure analog inputs for the PIC18F252 microcontroller. Both registers are shown in Figure 5.5.

Each of these two registers contains important bits essential for the proper configuration and operation of the ADC module on

ADCON0

R/W-0	R/W-0	R/W-0	R/W-0	R/W-0	R/W-0	U-0	R/W-0
ADCS1	ADCS0	CHS2	CHS1	CHS0	GO/DONE	—	ADON

bit 7   bit 0

ADCON1

R/W-0	R/W-0	U-0	U-0	R/W-0	R/W-0	R/W-0	R/W-0
ADFM	ADCS2	—	—	PCFG3	PCFG2	PCFG1	PCFG0

bit 7   bit 0

Legend:
R = Readable bit   W = Writable bit   U = Unimplemented bit, read as '0'
- n = Value at POR   '1' = Bit is set   '0' = Bit is cleared   x = Bit is unknown

Figure 5.5 ADCON0 and ADCON1 registers used to configure the PIC18F252s ADC module

© Microchip Technology Inc. Reproduced with permission.

the PIC. The register ADCON0 bits are:

- *Bit 0 – ADON –* used to turn on the ADC by setting this bit to 1. If it is set to 0 (reset value) the ADC is switched off and PIC consumes less power during its operation.
- *Bit 2 – GO/DONE –* used to start the actual conversion. When 1 is written to this bit (by program) conversion starts; once the value of this bit goes back to 0, the conversion is complete.

The result of the ADC conversion can be accessed via two other registers, ADRESH and ANDRESL. The analog-to-digital converter on this PIC is a 10-bit device; the result of the conversion of the analog signal is a 10-bit binary number. Since both ADRESH and ADRESL are 8-bit registers they need to be combined in order to obtain a 10-bit result of conversion. ADRESH and ADRESL registers can be combined in two ways according to the position of the conversion result in those 16 bits. The 10-bit result can be 'left justified' with the eight most significant bits stored in ADRESH and two less significant bits of the conversion result stored in bits 7 and 6 of register ADRESL. 'Right justification' for the converted result implies that the two most significant bits of the result are stored in bits 1 and 0 of register ADRESH with the other eight bits filling in the entire ADRESL register. This is shown in Table 5.1, where we used symbolic binary number 'abcdefghij' to represent the result of

Table 5.1 Left and right justification after AD conversion

ADFM	Justification	ADRESH	ADRESL
0	Left	`abcd efgh`	`ij00 0000`
1	Right	`0000 00ab`	`cdef ghij`

conversion, stored in ADRESH and ADRESL. The justification of conversion results is controlled by bit 7 from the ADCON1 register (ADFM) as detailed in Table 5.1.

Why would you want to use left justification in your programs? If an 8-bit conversion result is sufficient in an application it can be obtained from a 10-bit result by discarding the two least-significant conversion bits. This is easily accomplished by left-justifying the conversion result and reading the ADRESH byte only.

- *Bits 3, 4 and 5 – CHS0, CHS1 and CHS2 –* analog channel select bits used to select between 5 input channels (channel 1 – pin AN0, channel 2 – pin AN1... channel 5 – pin AN4) in the following way:

    CHS2, CHS1, CHS0 = 0, 0, 0 = channel 0 = pin AN0
    CHS2, CHS1, CHS0 = 0, 0, 1 = channel 0 = pin AN1
    CHS2, CHS1, CHS0 = 0, 1, 0 = channel 0 = pin AN2
    CHS2, CHS1, CHS0 = 0, 1, 1 = channel 0 = pin AN3
    CHS2, CHS1, CHS0 = 1, 0, 0 = channel 0 = pin AN4

- *Bits 6 and 7 – ADCS1, ADCS0 –* conversion clock selection bits (F_{OSC} is system clock):

    ADCS1, ADCS0 = 0, 0 = $F_{OSC}/2$ (ADCS2 = 0)
    ADCS1, ADCS0 = 0, 1 = $F_{OSC}/8$ (ADCS2 = 0)
    ADCS1, ADCS0 = 1, 0 = $F_{OSC}/32$ (ADCS2 = 0)
    ADCS1, ADCS0 = 1, 1 = F_{RC} (clock derived from internal RC oscillation) (ADCS2 = 0)

The above clock rates assume that the value of bit 6 from register ADCON1 (ADCS2) is set to 0. Additional conversion clock rates can be achieved if this bit is set to 1:

    ADCS1, ADCS0 = 0, 0 = $F_{OSC}/4$ (ADCS2 = 1)
    ADCS1, ADCS0 = 0, 1 = $F_{OSC}/16$ (ADCS2 = 1)

Table 5.2   Configuration of digital and analog channels of Port A

PCFG <3.0>	AN7	AN6	AN5	AN4	AN3	AN2	AN1	AN0	V$_{REF+}$	V$_{REF-}$
0000	A	A	A	A	A	A	A	A	V$_{DD}$	V$_{SS}$
0001	A	A	A	A	V$_{REF+}$	A	A	A	AN3	V$_{SS}$
0010	D	D	D	A	A	A	A	A	V$_{DD}$	V$_{SS}$
0011	D	D	D	A	V$_{REF+}$	A	A	A	AN3	V$_{SS}$
0100	D	D	D	D	A	D	A	A	V$_{DD}$	V$_{SS}$
0101	D	D	D	D	V$_{REF+}$	D	A	A	AN3	V$_{SS}$
011x	D	D	D	D	D	D	D	D	—	—
1000	A	A	A	A	V$_{REF+}$	V$_{REF-}$	A	A	AN3	AN2
1001	D	D	A	A	A	A	A	A	V$_{DD}$	V$_{SS}$
1010	D	D	A	A	V$_{REF+}$	A	A	A	AN3	V$_{SS}$
1011	D	D	A	A	V$_{REF+}$	V$_{REF-}$	A	A	AN3	AN2
1100	D	D	D	A	V$_{REF+}$	V$_{REF-}$	A	A	AN3	AN2
1101	D	D	D	D	V$_{REF+}$	V$_{REF-}$	A	A	AN3	AN2
1110	D	D	D	D	D	D	D	A	V$_{DD}$	V$_{SS}$
1111	D	D	D	D	V$_{REF+}$	V$_{REF-}$	D	A	AN3	AN2

$$\text{ADCS1, ADCS0} = 1, 0 = F_{OSC}/64 \,(\text{ADCS2} = 1)$$
$$\text{ADCS1, ADCS0} = 1, 1 = F_{RC} \,(\text{clock derived from RC}$$
oscillation) (ADCS2 = 1)

Important bits from the register ADCON1 are:

- *Bits 0, 1, 2 and 3 – PCFG0, PCFG1, PCFG2 and PCFG3–* analog or digital mode selection bits. Those bits are used to select which pins on PORTA are going to be used as analog and which are going to be used as digital pins in the program, according to Table 5.2.
- *Bit 6 – ADCS2 –* additional conversion clock selection bit, already discussed.
- *Bit 7 – ADFM –* justification selection bit. If this bit is set to 0, the conversion result in registers ADRESH and ADRESL is left justified and if it is set to 1, the result is right justified, as explained in Table 5.1.

Taking into consideration the functionality of various registers from ADCON0 and ADCON1 the program to perform voltage measurement and preset level checking and indication is given in Listing 5.1.

```
 1    // this program reads the voltage from RA0
 2    // and sets the alarm on RC3, if the voltage level exceeds
 3    // the preset "alarm_level" value
 4
 5    #include <p18F252.h>
 6
 7    float getVoltage(void)
 8    {
 9        ADCON0bits.GO = 1;                 //start a conversion
10        while (ADCON0bits.GO == 1);        // wait for completion
11        return(((ADRESH * 256) + ADRESL) * 0.00489);
12    }
13
14    void main (void)
15    {
16        float alarm_level = 1.;
17
18        ADCON0 = 0x81;       // select input AN0, enable ADC
19        ADCON1 = 0x80;       // RA0 is analog, right justified ADC
                                  result
20        TRISA = 0x01;        // RA0 is input
21        TRISC = 0x00;        // portC is output
22        PORTC = 0x00;        // alarm off
23        while(1)
24        {
25            if (getVoltage() > alarm_level)
26                PORTC = 0x08;
27            else
28                PORTC = 0x00;
29        }
30    }
```

Listing 5.1 Voltage level checking program using ADC

The operation of the analog-to-digital converter is configured at the beginning of the main function. By initializing the register ADCON0 to hex value 0x81in line 18, analog channel 0 (AN0) is selected, because the value of all three channel selection bits is 0 (CHS2 = CHS1 = CHS0 = 0). At the same time $F_{OSC}/32$ conversion clock rate is selected (ADCS1 = 1, ADCS0 = 0) and the conversion module is enabled (ADON = 1). By loading ADCON1 with hex value 0x80 in line 19, the lowest five pins of PORTA are all set into an analog mode of operation, according to Table 5.2. This program only uses channel 0, so the voltage source has to be connected to pin 2 – AN0. Bit 7 of ADCON1 is set to 1, so the result of the conversion, once the conversion is completed, will be right justified and available in registers

ADRESH and ADRESL. Line 20 is used to configure the same pin as an input pin, by setting the least significant bit of the TRISA register to 1. The other bits of this register are set to 0, so the other pins of PORTA are configured to be output pins, although they are not used in this program. Similarly, line 21 configures all pins of PORTC as output pins by writing hex value 0x00 to the TRISC register. Only pin 0 of PORTC is used in this program. Initially, it is set to 0 at the start of the program, in line 22. The content of registers ADCON0 and ADCON1 after configuration is shown in Figure 5.6 with some additional explanation.

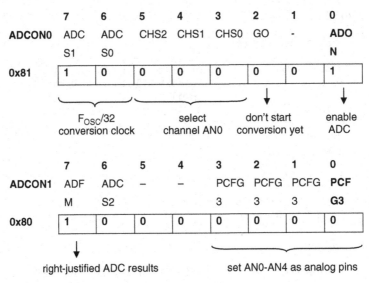

Figure 5.6 ADC configuration via ADCON0 and ADCON1 registers

The rest of the main part of the program is executed continuously during the program operation. The getVoltage() function is called to read the voltage from channel 0, convert it to a float value and compare it with the preset 'alarm_level'. If the obtained voltage value is greater than 'alarm_level', pin 0 of PORTC is set to 1; otherwise, it is set to 0.

The most important part of the program – analog to digital conversion – is started and completed in the function getVoltage(). The conversion is initiated by setting the bit GO/DONE from the ADCON0 register to 1 in line 9. It is interesting to notice the method of addressing bit GO/DONE from this register, as an element of the structure ADCON0bits – ADCON0bits.GO == 1. In fact, all of the bits from the ADCON0 register are defined in the pic18F252.h header file as elements of the structure ADCON0bits and can only be accessed in a similar way. In addition, all eight bits from this register can be addressed as a group by writing to or reading from ADCON0 as this register has also been defined as a character – a group of eight bits in the same header file. This is done in line 18 of this program.

In line 10, bit GO/DONE is continuously polled – its value checked. Initially at the start of the conversion the value of this bit is 1, as set in line 9 of the program, but it changes to 0 once the conversion is completed. The result of the conversion is available, in registers ADRESH and ADRESL. Knowing that the result is right justified – that the two most significant bits of the result are written to bits 0 and 1 of the ADRESH register while the rest of the result, the least significant 8 bits, is stored in the ADRESL register – the combination of those two registers is obtained by multiplying the value of register ADRESH by 256 and adding it to the value of register ADRESL. In this way, the two least significant bits of register ADRESH are shifted left by eight places and added to eight bits from register ADRESL. Finally the result of this operation needs to be scaled so that it corresponds to the voltage value from the operating range of the analog-to-digital converter. Assuming a supply voltage of 5 V, the operating range of the converter is 0 V to 5 V. The conversion factor and the voltage value of the least significant result bit is easily obtained using the equation below:

$$\text{conversion factor} = \frac{\text{ADC operating range}}{2^{10} - 1} = \frac{5\,\text{V} - 0\,\text{V}}{1024 - 1}$$

$$= \frac{5}{1023} = 0.00488756 \approx 0.00489$$

Therefore, after the multiplication of the combined ADRESH/ADRESL values by the conversion factor of 0.00489 (float), the actual voltage level is obtained and returned to the main program. In the main program the returned float value is compared with the preset 'alarm_level' in line 25 and the corresponding action is taken (lines 26 or 28).

The program can now be compiled and downloaded to the PIC chip to be tested in operation. For a relatively simple program like the one discussed in this section, the correct operation is the most likely outcome. For more complex programs involving analog signals, it might be more difficult to discover the problems and errors in the program during the program operation. The 'Stimulus' feature of the MPLAB development environment and the MPLAB SIM debugger can be useful in these situations. To demonstrate some features of this tool, we will show how to use it in order to test the operation of our program by simulating the program and providing stimulus to the AN0 pin. You should already be familiar with MPLAB SIM and some 'Stimulus' features as they were introduced and discussed in Chapter 1. Here we provide some more details on how to use this feature to simulate analog input to the microcontroller in order to test the operation of our voltage-level checking program.

To do this we first need to go through the usual process of writing and compiling the program in an MPLAB environment. As we are going to use the PIC18F252 device for this project, when specifying the language toolsuite in the Project Wizard dialog box we need to select the Microchip 18 Toolsuite and MPLAB C18 C Compiler options as shown in Figure 5.7.

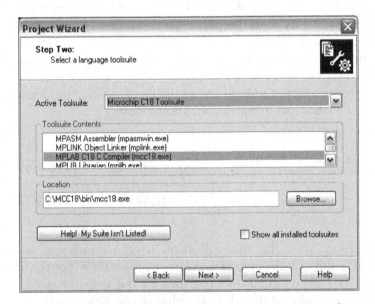

Figure 5.7 Selecting language tools in Project Wizard

© Microchip Technology Inc. Reproduced with permission.

Once the project is created, you might also need to specify an appropriate library path with C18 libraries, for example C:\MCC18 \lib or similar, depending on the location you have selected during the C18 compiler installation. The dialogue box is shown in Figure 5.8. This box is opened by selecting BUILD OPTIONS – PROJECT from the PROJECT menu in MPLAB.

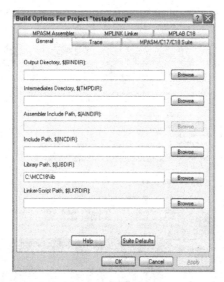

Figure 5.8 Library path specification for the PIC18 project

© Microchip Technology Inc. Reproduced with permission.

An appropriate linker file, 18f252.lkr for PIC18F252 residing in the MCC18/lkr subdirectory, should be added to the project.

After successful compilation of the program we can simulate the program in the MPLAB environment. To do that we need to select the MPLAB SIM tool from the DEBUGGER-SELECT TOOL menu of MPLAB. A new toolbar will be added to the MPLAB toolbar as indicated in Figure 1.18 in Chapter 1.

In order to simulate the operation of the ADC we need to select the STIMULUS option from the DEBUGGER menu. To set up the stimulus, select the NEW WORKBOOK option from the STIMULUS menu as shown in Figure 5.9. A new stimulus dialogue box will now open in MPLAB.

In Chapter 1 we discussed how to set up asynchronous and sequential-type stimuli appropriately in MPLAB. Here we show how to apply the cyclic-type stimulus, as it is the most appropriate stimulus type for the voltage level checking program given in Listing 5.1.

To apply the cyclic stimulus we need to generate a data file containing the data we want to apply to the RA0 pin of PORTA. Data are contained in text file 'wave1.txt', generated using the Notepad program. The content of part of this file is shown in Figure 5.10. This file can be found and downloaded from this book's web site. Data from the file are also graphically presented in the same figure. Note the range of data from the file 'wave1.txt' – 0 to 1000. The associated analog voltage

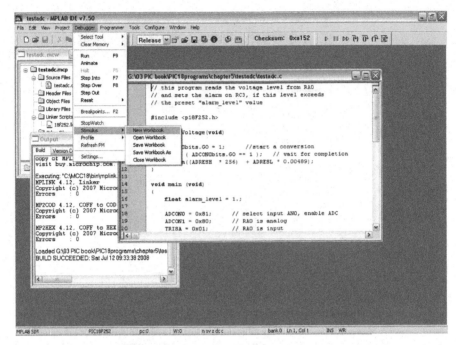

Figure 5.9 Opening new stimulus workbook in MPLAB

© Microchip Technology Inc. Reproduced with permission.

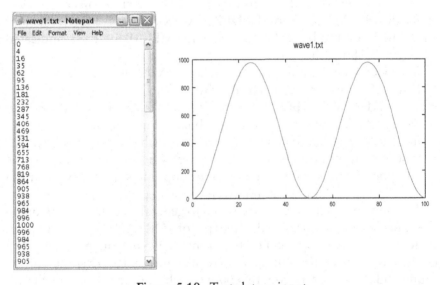

Figure 5.10 Test data – input

© Microchip Technology Inc.

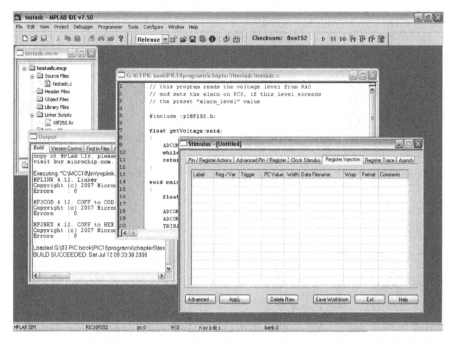

Figure 5.11 Preparing to configure the Register Injection type stimulus
in MPLAB

© Microchip Technology Inc. Reproduced with permission.

range, 0 V–4.89 V, obtained using the conversion factor of 0.00489, corresponds roughly to the full PIC ADC input voltage range.

To set up the ADC conversion register to receive its input from the generated data file we need to select the Register Injection tab in the opened Stimulus window as shown in Figure 5.11.

A default set of options is activated by clicking with the mouse in the first blank row of the Register Injection table. The register to be injected with stimulus is ADRESL by default and this selection should not be changed. The ADRESH register does not need to be selected; it will be filled in automatically in combination with the selected ADRESL register when needed. To select the data file, click on the field labelled 'Browse for Filename', find the directory containing the 'wave1.txt' file, select that file and click on the OPEN button in the file selection dialogue box. Change the selected file format from hexadecimal (Hex) to ASCII decimal (Dec) by clicking on the cell in the Format column of the Register Injection table. The numbers from our 'wave1.txt' data file will be interpreted as decimal values. The corresponding MPLAB window with selected options for Register Injection is shown in

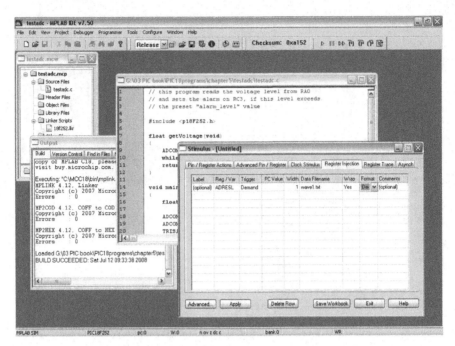

Figure 5.12 Register Injection options specified in the Stimulus dialogue box

© Microchip Technology Inc. Reproduced with permission.

Figure 5.12. This completes the basic stimulus input configuration. At the end of the file, MPLAB SIM will rewind and start over from the beginning of the file. The values from the file will be sent to the ADC register on demand – each time the ADC conversion is triggered by our program the next value from the file will be read into the ADRESL and ADRESH registers. To confirm and apply these settings click on the Apply button from the Stimulus window.

In order to record the output generated by our program, we also need to use the Register Trace tab in the Stimulus dialogue box. The new table should appear as depicted in Figure 5.13.

Here we need to set PORTC to be logged on demand; in other words, each time our program outputs the value to PORTC this value will be recorded as a decimal number in the selected output data file. We therefore need to generate another text file using the Notepad program and select that file when filling in this table. The file we generated for this experiment is called 'wave1_out.txt'. No data need to be generated for this file. It can be left empty to be filled by MPLAB during program execution, so the reader could easily generate this file using the Notepad program. By clicking on the first blank row of the Register Trace table a set of default options is activated. Those options need to be modified

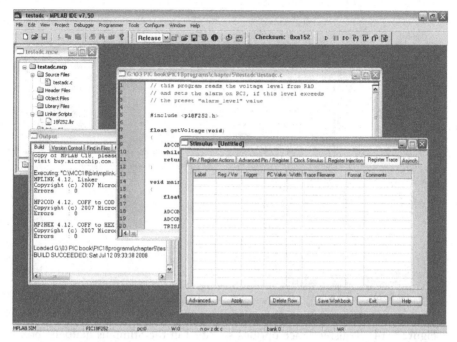

Figure 5.13 Preparing to configure Register Trace in MPLAB

© Microchip Technology Inc. Reproduced with permission.

according to Figure 5.14. The register to be traced is PORTC; the file holding trace information is 'wave1_out.txt' and the data format is decimal. The data trace will be triggered on demand each time data are written to PORTC. It will also be recorded in the specified file. Finally, to apply these settings click on the Apply button in the Stimulus window.

We are now ready to run and test our program. To do this, select then RUN or ANIMATE option from DEBUGGER menu.

After stopping the application, the results of the program – a set of values ouptut to PORTC – are stored in the specified file 'wave1_out. txt'. We can open this file to look at those values or, if we have access to a suitable software package (Matlab, Excel or similar), plot the data from this file. Part of the file 'wave1_out.txt' and its graphical representation are shown in Figure 5.15; 00 values in this data file indicate the 'measured' voltage level below the preset 'alarm_level' of 1 V. The LED attached to pin 3 of PORTC will be OFF at those times. The values of 08 indicate that pin 3 of PORTC has changed the state to 1 to indicate the input voltage level above 1 V. Comparing this waveform with input voltage values we notice that when the voltage level exceeds 1 V the PORTC signal changes from the 00 (pin 3 OFF) to the 08 (pin RC3 ON) state. This confirms the correct operation of our program.

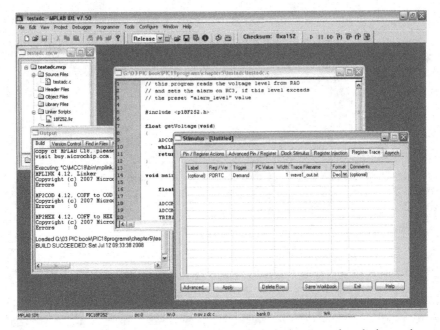

Figure 5.14 Register Trace options specified in the Stimulus dialogue box
© Microchip Technology Inc. Reproduced with permission.

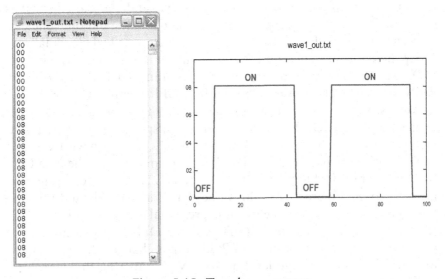

Figure 5.15 Test data – output
©Microchip Technology Inc.

5.4 Level Indicator with Timing

In this section we will modify our program slightly, in order to provide some
timing for its action of reading the input voltage. Our code will keep
checking the level of input voltage from analog channel 0 but it will do so in
regular time intervals. To achieve this, we will add an additional time delay
function to our program. The modified program is shown in Listing 5.2 and
the time-delay generating function wait() is defined between lines 9 and 13.

```
1   // this program reads the voltage level from RA0
2   // and sets the alarm on RC3, if this level exceeds
3   // the preset "alarm_level" value
4   // timing for the level checks is provided
5   // using simple software delay function
6
7   #include <p18F252.h>
8
9   void wait(void)
10  {
11    unsigned int x;
12    for (x=0; x<2650; x++) {}         // time delay
13  }
14
15  float getVoltage(void)
16  {
17  ADCON0bits.GO = 1;                  //start a conversion
18  while (ADCON0bits.GO == 1);         // wait for completion
19  return(((ADRESH * 256) + ADRESL) * 0.00489);
20  }
21
22  void main (void)
23  {
24    float alarm_level = 1.;
25
26  ADCON0 = 0x81;                      // select input AN0,
                                           enable ADC
27  ADCON1 = 0x80;                      // RA0 is analog
28  TRISA = 0x01;                       // RA0 is input
29  TRISC = 0x00;                       // portC is output
30  PORTC = 0x00;                       // alarm off
31  while(1)
32  {
33      if (getVoltage() > alarm_level)
34        PORTC = 0x08;
35      else
36        PORTC = 0x00;
37      wait();
38  }
39  }
```

Listing 5.2 Voltage level checking program with timing

To achieve the external voltage sampling time of 10 ms (100 Hz sampling frequency) the count limit for the 'for' loop in the wait() function needs to be adjusted. We have to run the program and measure the delay achieved with the wait() function using the Stopwatch feature of MPLAB SIM. This process has already been explained in Chapter 1 but we will repeat the main steps in this section.

A breakpoint should be set at a suitable place in the program, for example at line 37 where the wait() function is called from the main program. After reaching the breakpoint for the first time, the STOPWATCH option from the DEBUGGER menu can be used to time the execution of the program and the program should be run again (RUN from the DEBUGGER menu or F9). After another program break at the same breakpoint, the Stopwatch dialogue box indicates the elapsed time between two successive program breaks. This is the voltage sampling time. This process needs to be repeated several times for different values in the 'for' loop of the wait() function until a delay of approximately 10 ms is achieved. Every time the value in the 'for' loop is changed, the program is modified so it needs to be built again. For the value of 2650 in the 'for' loop and the processor frequency of 16 MHz, execution time is 10.0365. The MPLAB screenshot is shown in Figure 5.16. You can try this exercise for different processor frequencies.

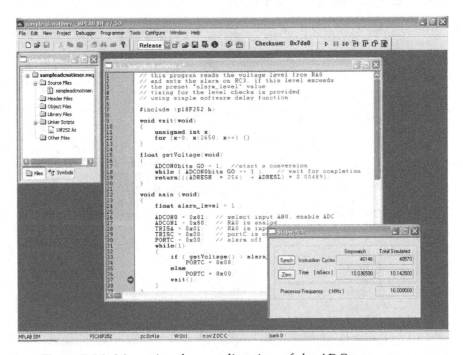

Figure 5.16 Measuring the sampling time of the ADC test program

The program execution-delaying method we have used in this example is relatively simple to implement and readers might notice that a similar approach has already been used for a number of programs in this book where some kind of delay had to be generated in order to achieve correct program operation. Still, this method has a couple of disadvantages. Manual adjustment of the time delay, using the trial-and-error approach explained in this section, is not the most accurate solution. The delay achieved in this way was inaccurate – 10.0365 ms instead of the desired 10 ms. The error is not too big but might be undesirable for more complex applications and external devices where more accurate timing is required.

The other important disadvantage of this program is its inefficiency. Using the Stopwatch feature we can test the timing of this program further to establish that almost half of the 10.0365 ms time needed for the program to reach the same breakpoint again was spent in the wait() function. During this period the program was not doing anything useful apart from simply waiting for the 'for' loop to finish. Obviously, this time was wasted and could have been used by our program to do some more useful things. This was not a problem when a simple program like the one shown in Listing 5.2 was needed. Here, only a simple comparison of the sampled voltage value was made and the appropriate value sent to PORTC in response. Execution time of the program and the use of instruction cycles will certainly become a more important issue when more complex projects are attempted. In these cases PIC will be required to do more and more processing with the sampled voltage value before the new sample is obtained from the analog channel. To achieve this, the more efficient approach to the program timing problem needs to be adopted. This can be achieved using the timer interrupt technique explained in the next section.

5.5 Level Indicator with Better Timing – Timer Interrupts

An interrupt is an event in program execution that forces a call to a special type of subroutine called an interrupt service routine (ISR). Interrupts are therefore routines but they are different from ordinary subroutines (when using assembler) or functions (when using C) as they are called by microcontroller hardware rather than a program/software itself. A good example of a software-called routine/function is the wait() function from the previous program. It was called by the program every time the execution of our program reached line 37 in Listing 5.2. The timing interrupt routine on the other hand will only be called when the end of the predetermined time interval is reached. This is achieved by using timers in the PIC18F252. The PIC18F252 has four timers – Timer 0, Timer 1, Timer 2 and Timer 3. All of these timers are similar in operation but there are some notable differences. To explain the interrupt feature of PIC as simply as possible in this first encounter with PIC timers we will discuss just one of these devices – Timer 2.

Physically, an 8-bit timer is a register whose value is continually increasing to 255 and then it starts all over again: 0, 1, 2, 3, 4 . . . 255,0, 1, 2, 3 . . . and so forth. As this incrementing is done in the background of normal microcontroller operation and according to a clock signal, it can be used to provide accurate timing where and when needed elsewhere in the program. This is usually done by starting the required action in the program at the moment of the timer overflow. The timer overflow is the moment when the timer reaches its full count and rolls over to zero to start another sequence of counting. Timer overflow is one of the events that can cause a call to interrupt a service routine during the program execution. To be able to do this, the PIC interrupt feature must be enabled (it is usually disabled by default when the PIC is switched on) and the timer must be initialized to achieve the desired overflow rate. This is done by configuring T2CON (Timer 2 Control Register), INTCON and PIE1 registers for PIC18F252. Those registers, along with one more interrupt-related register – PIR1, are shown in Figure 5.17. Some of the bits from those registers are related to Timer 2 operation and are discussed in the following section.

To enable Timer 2 interrupts, the following four bits need to be set to 1:

- *TMR2ON* – this bit in the T2CON register switches Timer 2 on (when set to 1) or off (when set to 0).
- *GIE/GIEH* – this bit in the INTCON register globally enables all enabled interrupts.
- *PEIE/GIEL* – this bit in the INTCON register enables all peripheral type interrupts (Timer 2 interrupt is one of those).
- *TMR2IE* – this bit from the PIE1 register additionally enables Timer 2 interrupts.

It is interesting to note that not one but three bits are used to enable Timer 2 interrupt on PIC18F252. Although only one of those bits relates directly to Timer 2 (TMR2IE), the other two interrupt enable bits are there as 'additional insurance'. A group of interrupts can be collectively switched off by setting the PEIE/GIEL bit to 0 while all interrupts, although individually enabled, can be switched off with a simple action of setting the GIE/GIEH bit to 0. This usually simplifies programs where

T2CON: TIMER2 CONTROL REGISTER

U-0	R/W-0	R/W-0	R/W-0	R/W-0	R/W-0	R/W-0	R/W-0
—	TOUTPS3	TOUTPS2	TOUTPS1	TOUTPS0	TMR2ON	T2CKPS1	T2CKPS0

bit 7 — bit 0

INTCON REGISTER

R/W-0	R/W-0	R/W-0	R/W-0	R/W-0	R/W-0	R/W-0	R/W-x
GIE/GIEH	PEIE/GIEL	TMR0IE	INT0IE	RBIE	TMR0IF	INT0IF	RBIF

bit 7 — bit 0

PIE1: PERIPHERAL INTERRUPT ENABLE REGISTER 1

R/W-0	R/W-0	R/W-0	R/W-0	R/W-0	R/W-0	R/W-0	R/W-0
PSPIE[1]	ADIE	RCIE	TXIE	SSPIE	CCP1IE	TMR2IE	TMR1IE

bit 7 — bit 0

PIR1: PERIPHERAL INTERRUPT REQUEST (FLAG) REGISTER 1

R/W-0	R/W-0	R-0	R-0	R/W-0	R/W-0	R/W-0	R/W-0
PSPIF[1]	ADIF	RCIF	TXIF	SSPIF	CCP1IF	TMR2IF	TMR1IF

bit 7 — bit 0

Legend:			
R = Readable bit	W = Writable bit	U = Unimplemented bit, read as '0'	
- n = Value at POR	'1' = Bit is set	'0' = Bit is cleared	x = Bit is unknown

Figure 5.17 Timer 2 configuration registers

© Microchip Technology Inc. Reproduced with permission.

a number of interrupts need to be controlled and enabled or disabled at various locations in the program.

How does the PIC know about the occurrence of Timer 2, or indeed any other type of interrupt? A special function register, PIR, contains some important bits used to signal or 'flag' the interrupt occurrence. Bit **TMR2IF** is a Timer 2 interrupt flag. When Timer 2 reaches its full count this flag is set (by processor hardware) to 1 and the program operation is redirected to the interrupt service routine – a function written to service a particular interrupt. It is wise to clear this flag at the end of the interrupt service routine in order to prepare for the possible occurrence of the same type interrupt at some later time during the program operation.

In order to set the period for timer interrupts we need to combine the other six bits from the T2CON register with the value of 8-bit Timer 2 period register PR2. A simple formula to calculate the Timer 2 interrupt rate is:

$$\text{Time 2 interrupt rate} = A \times B \times C \times T_{ic}$$

In the above equation:

- A is related to the content of bits 3–6 from the T2CON register (TUOTPS0–TOUTPS3)
- B is related to values of bits 0 and 1 from the same register (T2CKPS1 and T2CKPS0)
- C is the value determined by the content of period register PR2, and
- T_{ic} is the instruction cycle duration for the particular device.

Tables 5.3 and 5.4 define the values of A and B with respect to those bits.

Table 5.3 Configuration bits A for Timer 2 interrupt rate adjustment

TOUTPS3	TOUTPS2	TOUTPS1	TOUTPS0	A
0	0	0	0	1
0	0	0	1	2
0	0	1	0	3
0	0	1	1	4
0	1	0	0	5
0	1	0	1	6
0	1	1	0	7
0	1	1	1	8
1	0	0	0	9
1	0	0	1	10
1	0	1	0	11
1	0	1	1	12
1	1	0	0	13
1	1	0	1	14
1	1	1	0	15
1	1	1	1	16

Table 5.4 Configuration bits B for Timer 2 interrupt rate adjustment

T2CKPS1	T2CKPS0	B
0	0	1
0	1	4
1	0	16
1	1	16

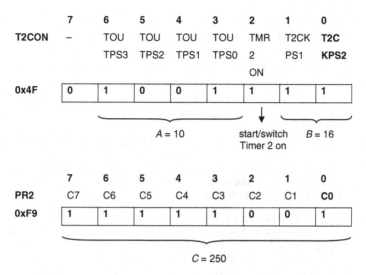

Figure 5.18 Interrupt rate configuration for Timer 2

For a 16 MHz clock, the duration of the instruction cycle is $T_{ic} = \frac{4}{16.10^6} = 0.25\,\mu s$ so 10 ms interrupt timing can be achieved with $A = 10$, $B = 16$ and $C = 250$ values, because

$$
\begin{aligned}
\text{Timer 2 interrupt rate} \ &= \ A \times B \times C \times T_{ic} \\
&= \ 10 \times 250 \times 16 \times 0.25 \times 10^{-6}\,\text{s} \\
&= \ 10^{-2}\,\text{s} \\
&= \ 10\,\text{ms}
\end{aligned}
$$

Required loadings of registers T2CON and PR2 are shown in Figure 5.18.

In addition to this, relevant bits from registers INTCON and PIE1 need to be set at the beginning of the program. The whole program is given in Listing 5.3.

```
1    // this program reads the voltage level from RA0
2    // and sets the alarm on RC3, if this level exceeds
3    // the preset &ldquo;alarm_level&rdquo; value
4    // sampling rate for the input voltage signal
5    // is set to 100 Hz using Timer2 interrupt
6
7    #include <p18F252.h>
8
9    float alarm_level = 1.;
10
```

Listing 5.3 Voltage level checking program using Timer 2 interrupt

```
11  void timer_isr (void);
12  float getVoltage(void);
13
14  #pragma code int_vector=0x08 // start of ISR part 1
15  void high_interrupt (void)
16  {
17    _asm GOTO timer_isr _endasm   // just redirects to part 2
18  }
19
20  #pragma code                    // end of ISR part 1
21
22  #pragma interrupt timer_isr   // start ISR part 2
23  // this is where main ISR work is done
24    void timer_isr (void)
25  {
26      if (getVoltage() > alarm_level)
27          PORTC = 0x08;
28      else
29          PORTC = 0x00;
30    PIR1bits.TMR2IF = 0;
31  }
32
33    void Timer2_Init(void)
34  {
35      T2CON = 0x4F;              // configure timer to
36      PR2 = 0xF9;                // 10 ms sampling interval
37      PIE1bits.TMR2IE = 1;       // enable Timer 2 interrupts
38      INTCONbits.PEIE = 1;       // enable peripheral
                                    interrupts
39  }
40
41  float getVoltage(void)
42  {
43    ADCON0bits.GO = 1;     //start a conversion
44    while (ADCON0bits.GO == 1);   // wait for completion
45    return((ADRESH * 256 + ADRESL) * 0.00489);
46  }
47
48  void main (void)
49  {
50      float alarm_level = 1.;
51
52      ADCON0 = 0x81;             // select input AN0, enable ADC
53      ADCON1 = 0x80;             // RA0 is analog
54      TRISA = 0x01;              // RA0 is input
55      TRISC = 0x00;              // portC is output
56      PORTC = 0x00;              // alarm off
57      Timer2_Init();             // set the timer
```

Listing 5.3 (*continued*)

```
58      INTCONbits.GIE = 1;             // enable interrupts
59
60      while(1);                 // wait for interrupts
61    }
```

Listing 5.3 *(continued)*

The main() function of this program is similar to the main() function of the previous program but an extra two lines, 57 and 58, are added. In line 57, a function to initialize Timer 2 – Timer2_Init() – is called and in line 58 interrupts have been enabled by setting the GIE bit from the INTCON register to 1. This bit is accessed as a member of the INTCONbits structure, similarly to the GO/DONE bit from the ADCON0 register in section 5.3. Timer2_Init() function is defined between lines 33 and 39 of the program. It loads the T2CON and PR2 registers with hexadecimal values 0x4F and 0xF9 to achieve appropriate timing using Timer 2. Lines 37 and 38 fully enable Timer 2 interrupts by setting bits TMR2IE from the PIE1 register and PEIE from the INTCON register to 1. These two bits are also accessed individually as members of two structures defined in the p18F252.h header file for this PIC model.

The function to initiate and read the result of analog-to-digital conversion, getVoltage(), is the function already used in the program in listing 5.2, so it does not need further explanation.

Most of the work done in this program is defined within function timer_isr() – the Timer 2 interrupt service routine. This function is defined between lines 22 and 31. As we want to achieve a 100 Hz sampling rate for analog channel 0, ADC is read – function getVoltage() is called every time an interrupt occurs. This function returns the current value of the sampled voltage from the input pin so on line 26 the program calls this function and compares the returned value with the preset voltage value 'alarm_level'. The remainder of the timer_isr() function follows the same logic as the main part of the previous voltage level checking program. Those are lines 33–36 from Listing 5.2.

Line 30 of the function timer_isr() clears the interrupt flag to enable recognition of the next Timer 2 interrupt.

The rest of the program is dedicated to the proper setting of the interrupt service routine. Unlike ordinary subroutines in assembler language or functions in C language interrupt service, routines need to be placed at predefined memory locations. For most of the PIC processors this location is 08 in hexadecimal format. Another location, 18h, is also available in the PIC18F252 microcontroller when more than one interrupt source is enabled and expected to occur in the program. As our program only has one interrupt source, Timer 2 overflow, we have to store the first ISR instruction in location 0x08. This is done using C directive #pragma in line 14. The ISRs can grow quite big, so for more complex programs and projects a common approach is to use the 0x08 memory location to redirect program execution to the remaining part of the ISR elsewhere in PIC memory. Here, we use 'inline assembly code' to redirect program execution to already discussed function timer_isr(). This redirection is done in line 17 of our program – program execution is redirected (GOTO) to function timer_isr. When Timer 2 overflows, causing the interrupt, normal program execution is suspended and the program jumps to address 0x08. Here, it is redirected again to continue by executing the timer_isr() function located elsewhere in the PIC program memory. Once this function is completed, the program goes back to the main part and resumes normal operation. In this case, the program goes back to line 60 'while(1)' – doing nothing and just waiting for the next interrupt to occur. The #pragma code directive on line 20 indicates the end of the first part of the ISR, which was announced with the first #pragma directive in line 14. The third #pragma directive, on line 22, indicates the start of the second part of the ISR, function timer_isr().

A good exercise for a keen reader would be to simulate the execution of this program before implementing it in hardware. By using the Stopwatch feature of the MPLAB SIM debugger, the exact sampling frequency (the time between two conversions) can be measured. It should be 10 ms or very close to this value.

Another useful exercise to try with this program is to change the loading of the Timer 2 registers T2CON and PR2 in order to change the rate of sampling and converting the applied analog voltage. From the new loadings of registers T2CON and PR2 the reader can calculate the sampling rate using the equation given earlier in this section. Alternatively the reader can determine the values to be loaded in those two registers in order to achieve the particular sampling rate. The program can then be simulated and the sampling rate can be measured to check the accuracy of calculation.

5.6 Talkthrough Program with Adjustable Sampling Rate

The final program in this chapter is the so-called 'talkthrough'-type program. This program does not do much but can be very useful if more complex programs are to be developed. It samples the analog channel

0 and sends the reading to PORTB. A network of resistors attached to PORTB converts the 8-bit digital value to analog signal. The circuit diagram for this project is given in Figure 5.4, and the corresponding program code is given in Listing 5.4. The program is very similar to the program discussed previously and as such does not require much explanation. The obvious difference between these two programs is that the level checking performed in the program from the previous section has been left out. Instead, the eight most significant bits from analog to digital conversion are simply copied to PORTB, performing the 'talkthrough' function of the program. This causes some accuracy loss as the full conversion result occupies 10 bits in registers ADRESH and ADRESL. This can be prevented by building a more complex resistor network and connecting it to eight pins of PORTB and two pins of PORTC. This is left as an exercise for the reader.

The sampling rate in this program has also been increased to allow the system to cope with higher frequency signals. This makes the program a good starting point for the processing of some audio signals. A related project, implementing some simple digital filters, is described in the next chapter of this book. The reader can try to calculate the sampling rate of the program given in Listing 5.4 and check the result by simulating the program using the MPLAB SIM debugger.

```
1    // this program reads the voltage level from RA0
2    // and outputs the corresponding 8 bit digital
3    // value to PORTB discarding the two least significant
4    // bits of the conversion result in the process
5
6    #include <p18F252.h>
7    #include <timers.h>
8
9    // Global Variables
10   unsigned char i = 0;
11
12   void timer_isr (void);
13   char getVoltage(void);
14
15   #pragma code int_vector=0x08  // start of ISR part 1
16   void low_interrupt (void)
17   {
18   _asm GOTO timer_isr _endasm   // just redirects to part 2
19   }
20
21   #pragma code                  // end of ISR part 1
22
```

Listing 5.4 Talkthrough program

```
23   #pragma interrupt timer_isr      // start ISR part 2
24   // this is where main ISR work is done
25   void timer_isr(void)
26   {
27   i = getVoltage();
28     PORTB = i;
29     PIR1bits.TMR2IF = 0;
30   }
31
32   void Timer2_Init(void)
33   {
34     T2CON = 0x4D;              // timer configuration
35     PR2 = 0x18;
36     PIE1bits.TMR2IE = 1;  // enable Timer 2 interrupts
37     INTCONbits.PEIE = 1;  // enable peripheral interrupts
38   }
39
40   char getVoltage(void)
41   {
42     ADCON0bits.GO = 1;          //start a conversion
43     while (ADCON0bits.GO == 1); // wait for completion
44     return((ADRESH * 64 + ADRESL / 4));  // convert to 8-bit
45   }
46
47   void main (void)
48   {
49
50     ADCON0 = 0x81;          // select input AN0, enable ADC
51     ADCON1 = 0x80;          // RA0 is analog
52     TRISA = 0x01;           // RA0 is input
53     TRISB = 0x00;           // portB is output
54     Timer2_Init();          // set the timer
55     INTCONbits.GIE = 1;     // enable interrupts
56
57   while(1);                 // wait for interrupts
58   }
```

Listing 5.4   (*continued*)

CHAPTER

6

Other PIC Projects

6.1 Introduction
6.2 Stepper Motor Controller using PIC12F675
6.3 DC Motor Controller using a PIC12F675
6.4 An Ultrasonic Measuring System using the PIC16F627A
6.5 Function Generator
6.6 Digital Filtering

6.1 Introduction

A number of PIC MCU-based projects will be described and implemented in this chapter. Circuit diagrams for most of the projects from the chapter will be provided but it will be left to readers to improve and further modify those circuits, designing their own PCB layouts where needed, using the knowledge acquired from projects in the previous chapters. C programs for all of the projects will be given but it will be left to readers to rewrite those programs using the assembler programming language if and when needed. Additional items of equipment are used for some of these projects. They will be described in some detail in the relevant sections of this chapter.

The PICs used in this chapter are the PIC12F675 and the PIC18F252.

6.2 Stepper Motor Controller using PIC12F675

In this project, a 12F675 PIC is used to control the speed and direction of a stepper motor. A PIC12F675 is used instead of a PIC12F629 because this chip provides optional analog inputs and we need to use all analog inputs for reading control buttons. The schematic diagram of this circuit is shown in Figure 6.1.

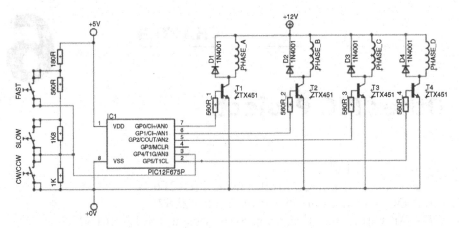

Figure 6.1 Stepper motor circuit

Let us explain the circuit used for the project here. The stepper motor used in this project requires a 12 V supply and can require up to 1 A per phase, which the PIC12F675 cannot supply. In fact, hardly any micro can supply this level of current to external circuitry. We have therefore used ZTX451 NPN bipolar transistors for switching stepper motor phases ON/OFF. Phases are coils, so during the switch OFF (breaking the current path of the inductive load) the current due to stored energy in the coil should have a path to continue flowing, otherwise generated back electromotive force (emf) can damage the transistors. Diodes are therefore placed across the coils in reverse to provide the path for the current flow. Four PIC I/O pins out of six are used to drive the stepper motor and we need to have Fast, Slow, Clockwise rotation (CW) and Counter clockwise rotation (CCW) control options, so we decided to use one of the remaining I/O pins as an analog input and use the resistor network shown on the left side of the circuit, which will provide the following switching function:

- When nothing is pressed: voltage on AN3 = $(5 \times 1000)/ (180 + 560 + 1800 + 1000) = 1.41$ V.
- When FAST is pressed: voltage on AN3 = $(5 \times 1000)/(180 + 1000) = 4.24$ V.
- When SLOW is pressed: voltage on AN3 = $(5 \times 1000)/(180 + 560 + 1000) = 2.87$ V.
- When CW/CCW is pressed: voltage on AN3 = $(5 \times 0)/(180 + 560 + 1800 + 1000) = 0$ V.

These voltage values are used to determine which of three push buttons is pressed if any. CW/CCW is used as a toggle button – every time this button is pressed, the motor direction is changed (between CW and CCW). The program used for this project is given in Listing 6.1.

```
1   #include <pic12f6x.h>
2
3   unsigned int state,count,CW,countMax;
4   unsigned char adcValue;
5
6   // Peripheral initialization function
7   void init(void){
8
9   /***** GPIO setup ****
10   * GPIO0,GPIO1,GPIO2,GPIO5 as outputs
11   * GPIO4(AN3) as analog input
12   */
13     ANS0  = 0;          // GPIO0 is digital
14     ANS1  = 0;          // GPIO1 is digital
15     ANS2  = 0;          // GPIO2 is digital
16     ANS3  = 1;          // GPIO4/AN3 is analog
17     TRIS0 = 0;          // set port pin 0 to output
18     TRIS1 = 0;          // set port pin 1 to output
19     TRIS2 = 0;          // set port pin 2 to output
20     TRIS5 = 0;          // set port pin 5 to output
21     TRIS4 = 1;          // set port pin 4 to input
22
23   /***** ADC setup ****
24   * ADC clock = 4MHz/16(4uS)
25   * AN3 selected
26   * results left justified (ADRESH=8 MSBs).
27   */
28       ADCS2=1; ADCS1=0; ADCS0=1;  // ADC clock = Fsoc/16
29         CHS1 =1; CHS0 =1;         // select channel 3
30         ADON =1;                  // turn on ADC
31         ADFM =0;                  // left justification
32
33   /***** Timer 0 setup ****
34   * prescale ratio = 1:256
35   * interrupt enabled
36   * clocked internally.
37   */
38       TMR0 = 205;
39       PS2 = 1;          // timer prescalar
40       PS1 = 0;          // set to
41       PS0 = 0;          // 1 : 32
42       PSA = 0;          // prescalar enabled
43       TOCS = 0;         // TMR0 set as timer
```

Listing 6.1 PIC12F675 Stepper motor program

```
44        TOIE = 1;        // enable timer0 interrupt
45     GIE = 1;    // Global interrupts enabled
46
47     state = 0;
48     count = 0;
49     CW = 1;
50     countMax = 1;
51     adcValue = 0;
52  }
53
54  // interrupt service routine ISR
55  void interrupt ISR(void){
56  /***** Timer 0 Code *****/
57     if((TOIE)&&(TOIF)){
58         count++;
59         if(count > countMax){
60             if(CW==1){
61                 switch(state) {
62                     case 0: state = 1; break;
63                     case 1: state = 2; break;
64                     case 2: state = 3; break;
65                     case 3: state = 0; break;
66                 }
67             }
68             else{
69                 switch(state) {
70                     case 0: state = 3; break;
71                     case 1: state = 0; break;
72                     case 2: state = 1; break;
73                     case 3: state = 2; break;
74                 }
75             }
76             count = 0;
77         }
78
79         switch(state) {
80           case 0: GPIO = 0x05; break;
81           case 1: GPIO = 0x21; break;
82           case 2: GPIO = 0x22; break;
83           case 3: GPIO = 0x06; break;
84                             }
85         TMR0 = 190;
86         TOIF=0;                  // clear event flag
87     }
88  }
89
90  void main(void){
91     init();
92     while(1){
93         GODONE = 1;          // start conversion
94         while(GODONE)        // wait for end of conversion
95             ;
96         if(ADRESH < 35){   // toggle direction when < 0.7v
```

Listing 6.1   (*continued*)

```
 97                    if(CW == 1) CW = 0;
 98                    else      CW = 1;
 99            }
100          if (ADRESH > 190)
101               countMax=1;      // Fast, when > 3.7v
102          else                  // slow, when > 2.1v and <3.2v
103              if((ADRESH>110)&&(ADRESH<170))countMax=20;
104         }
105      }
```

Listing 6.1 (*continued*)

The program starts by calling the 'init()' function (discussed in the next section) at line 91, which configures I/O pins and sets up the analog-to-digital converter (ADC) and the timer 0 interrupt. Then from line 93 to 103, the ADC is started and, when conversion is finished, depending on its value, variables CW (which determines the direction of rotation) and countMax (which determines Fast/Slow operation) are adjusted. As can be seen, the 'main()' function's task is to read the keys and adjust the variables.

The 'init()' function starts at line 7, and performs the following tasks:

1. Configuration of the I/O pins (lines 13–21).
2. Setting up the ADC (lines 28–31).
3. Timer 0 and interrupt initialization (lines 38–45).
4. Initialization of the variables used in the program (lines 47–51).

The interrupt service routine ISR (lines 55–88) is automatically called when an interrupt occurs. In this project only the timer 0 interrupt has been enabled, so the ISR is called when timer 0 times out. To drive a stepper motor, one should energize the phases in a certain sequence. There are many sequences available to drive a stepper motor but the one chosen in our program follows the pattern outlined below:

GP5	–	–	GP2	GP1	GP0	State
0			1	0	1	0
1			0	0	1	1
1			0	1	0	2
0			1	1	0	3

To generate the above sequence we have used state machine techniques and assigned each sequence a state. For example, state 2 would generate a 1010 pattern. If the sequence is in the order of 0, 1, 2 and 3, the motor would rotate clockwise and the reverse order would cause the motor to rotate counterclockwise. Lines 60–77 determine the direction and then set the state accordingly. For instance if the motor is running in a clockwise direction (CW) and the present state is 0, then the state at line 62 would be assigned a new value of 1. In this way the patterns are followed according to the value of CW. At lines 79–84, depending on which state it is, appropriate outputs are applied. To control the speed of the rotation a check is made at line 59 to determine whether the sequence that follows is to be run or not. If the condition is true the sequence would be run. Speed of motor rotation is determined by the value of variable 'countMax'. This variable is set in the main() part of the program according to the result of AD conversion, i.e. it reflects the value of the voltage supplied to pin AN3. When countMax $= 1$ (AN3 > 3.7V), the motor runs FAST, while countMax $= 20$ (AN3 < 3.2V) causes the motor to run SLOW.

To extend this project further, readers can consider adding more controls such as varying the speed, instead of just using FAST and SLOW, or even producing other patterns to drive the stepper motor.

6.3 DC Motor Controller using a PIC12F675

In this project, a 12F675 PIC is used to control the speed of a direct current (DC) motor. The control is implemented using the PWM (pulse width modulation) technique, where the supply voltage to the motor is switched ON and OFF repeatedly.

This approach is explained in Figure 6.2. The left diagram shows the power supply signal sent to the motor in order to achieve a lower speed. The pulse width is 25% of the total time T, so the average voltage is 25% of the DC voltage, V. The pulse width on the right diagram is 75% of the total time T so the average voltage is 75% of DC voltage. The average voltage supplied to the DC motor will, of course, determine the speed of

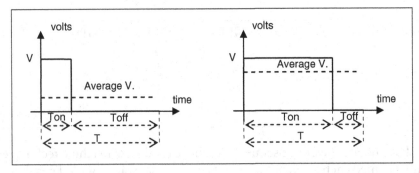

Figure 6.2 PWM techniques for varying average voltage level

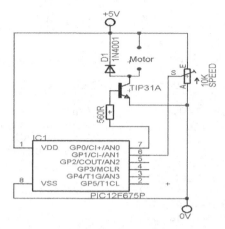

Figure 6.3 DC motor control circuit

its rotation, so by adjusting the pulse width of the signal we can actually control the speed of the DC motor.

The schematic diagram for this project is given in Figure 6.3. This is a simple circuit that uses very few external components.

For setting the speed, a 10k potentiometer is used and for switching the DC motor ON and OFF a TIP31A transistor is used to provide the higher current rating required by the motor. Again, as in the previous project (stepper motor), a reverse diode is added to the circuit to provide a path for motor inductive current. The program used in this project is given in Listing 6.2.

```
 1  #include <pic12f6x.h>
 2
 3  // Peripheral initialization function
 4  void init(void){
 5
 6  /***** GPIO setup ****
 7   * GPIO0,GPIO1,GPIO2,GPIO5 as outputs
 8   * GPIO4(AN3) as analog input
 9   */
10  ANS0 = 0;            // GPIO0 is digital
11  ANS1 = 1;            // GPIO1/AN1 is analog
12  ANS2 = 0;            // GPIO2 is digitalyyyyyyyyy
13  ANS3 = 0;            // GPIO4 is digital
14  TRIS0 = 0;           // set port pin 0 to output
15  TRIS1 = 1;           // set port pin 1 to input
16  TRIS2 = 0;           // set port pin 2 to output
17  TRIS5 = 0;           // set port pin 5 to output
```

Listing 6.2 PIC12F675 DC motor control program

```
18   TRIS4 = 0;                 // set port pin 4 to output
19
20   /***** ADC setup ****
21    * ADC clock = 4MHz/16
22    * AN3 selected
23    * results left justified (ADRESH=8 MSBs).
24    */
25   ADCS2=1; ADCS1=0; ADCS0=1; // ADC clock = Fsoc/16
26   CHS1 =0; CHS0 =1;          // select channel 1
27   ADON =1;                   // turn on ADC
28   ADFM =0;                   // left justification
29
30   /***** Timer 0 setup ****
31    * prescale ratio = 1:256
32    * interrupt enabled
33    * clocked internally.
34    */
35   TMR0 = 0;
36   PS2  = 1;                  // timer prescalar
37   PS1  = 0;                  // set to
38   PS0  = 0;                  // 1: 32
39   PSA  = 0;                  // prescalar enabled
40   T0CS = 0;                  // TMR0 set as timer
41   T0IE = 1;                  // enable timer0 interrupt
42
43   GIE  = 1;                  // Global interrupts enabled
44   }
45
46   void PWM_time(int pwm){    // pwm/255 delay
47       int i;
48       for(i=0;i<pwm;i++);
49   }
50
51    // interrupt service routine ISR
52   void interrupt ISR(void){
53        /***** Timer 0 Code *****/
54       if((T0IE)&&(T0IF)){
55           TMR0 = 90;          // 5.3 ms
56           T0IF = 0;           // clear event flag
57           GODONE = 1;         // start conversion
58           while(GODONE)       // wait for end of conversion
59             ;
60           GPIO0=1;            // motor ON
61           PWM_time(ADRESH);
62             GPIO0=0;          // motor OFF
63             PWM_time(255-ADRESH);
64       }
65   }
66
67   void main(void){
68       init();
69       while(1){}
70   }
```

Listing 6.2   (*continued*)

The program in Listing 6.2 starts with the main function consisting of only two lines. At line 68 the initialization function 'init()' is called and in line 69 the construct 'while(1)' is used in order to cause the program to stay on this line for the rest of the program execution. The 'init()' function is used to:

- set up the I/O pins (GPIO0 as digital output, GPIO1 as analog input)
- configure the ADC (select channel 1) and finally
- set up timer 0 and its interrupt.

In this program the period of the pulse applied to the transistor is fixed. To achieve this, the timer interrupt at line 52 is used to switch the transistor ON and OFF. When the timer interrupt occurs, the value of the ADC is read, which represents the speed required and hence determines the pulse width. At line 60 the transistor is switched ON and, after some time delay (depending on the 0–255 value read from the ADC), it is turned OFF again for another time period (255 – ADC value). The 'PWM_time()' function takes a variable and runs a 'for-loop' to produce the delay.

This project could be further developed to have a closed loop control for speed or position, with extra analog inputs available.

6.4 An Ultrasonic Measuring System using the PIC16F627A

In this project, a PIC16F627A is used for a distance-measurement system. This project is based on the LCD board in Chapter 3, with the addition of an ultrasonic module (module number). This module requires a power supply of +5 V, and works by applying a short pulse (minimum length of $10\,\mu s$) to its transmitter input. When this pulse is generated the transmitter produces a short ultrasonic burst and the receiver then produces a pulse with duration proportional to the distance of an object away from the transmitter. The length of this pulse needs to be measured to establish the distance of this object from our measurement system. From this measurement the distance can be calculated and displayed on the LCD. The schematic diagram of this circuit is shown in Figure 6.4.

In addition to the LCD interface (from Chapter 3) there are two extra push buttons included in this circuit as well as the ultrasonic module. One push button is used as a RESET button to reset the microcontroller

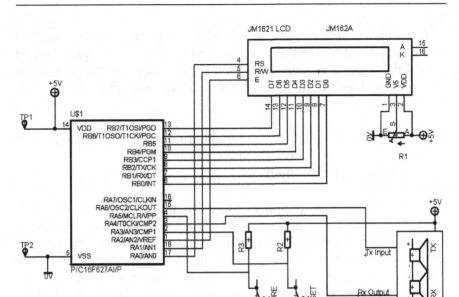

Figure 6.4 Distance-measurement circuit

(note that here the PIC is programmed with the MCLR pin enabled) and one push button is to initiate the distance measurement. These buttons are connected as active low, which means that, when pressed, they produce a logic 0 on the relevant connected I/O pin. For a more detailed explanation of the LCD interface the reader is referred to Chapter 3.

The program used in this project is given in Listing 6.3.

```
 1  //**********************************************
 2  // RB        7      6      5      4      3      2      1      0 *
 3  // Data Bus D7     D6     D5     D4     D3     D2     D1     D0 *
 4  // ------------------------------------------------------ *
 5  // RA        0      1      2                                  *
 6  // Controls RS     RW      E                                  *
 7  //**********************************************
 8  #include <pic16f62xa.h>
 9
10  #define RS       RA0
11  #define RW       RA1
12  #define E        RA2
13  #define NewLine                          0xC0
14  #define ClrDisp                          0x01
```

Listing 6.3 PIC 16F627A distance measurement program

```
15   #define TwoLine10dots            0x38
16   #define ScrOnCurOffBlinkOn       0x0F
17   #define IncCurDntMovDis          0x06
18   #define CurHome                  0x02
19
20   #define Measure     RA4
21   #define Tx          RA6
22   #define Rx          RA3
23
24   unsigned int    distanceInt;
25   unsigned char   distanceChar[5];
26
27   #define  NOP() asm("nop")   // 1 us delay for 4MHz clock
28
29   void delay_uS(unsigned int d) {
30     unsigned int j;
31      for (j = 0; j < d; j++) NOP();
32   }
33
34   void delay_mS(unsigned int D) {
35     unsigned int i;
36        for (i = 0; i < D; i++)
37           delay_uS(1000);
38   }
39
40   void sendCmd(unsigned char Cmd){// write text to LCD
41     PORTB = Cmd;                  // put char into RAM
42     RS = 0;                               // command mode
43     RW = 0;                           // write operation
44     delay_mS(1);
45     E = 1;
46     delay_mS(1);
47     E = 0;                            // send command to LCD
48     return;
49   }
50
51   void sendData(unsigned char Ch){// write text to LCD
52     PORTB= Ch;                    // put char into RAM
53      RS = 1;                                 // data mode
54   RW = 0;                            // write operation
55     delay_mS(1);
56     E = 1;
57     delay_mS(1);
58     E = 0;                              // send data to LCD
59     return;
60   }
61
62   void initializeLCD(void){
63      delay_mS(40);
64      sendCmd(TwoLine10dots);
```

Listing 6.3 *(continued)*

```
65      delay_mS(1);
66      sendCmd(ScrOnCurOffBlinkOn);
67      delay_mS(1);
68      sendCmd(ClrDisp);
69      delay_mS(1);
70      sendCmd(CurHome);
71      delay_mS(1);
72      sendCmd(0x06);
73   }
74
75   void intToascii(void){
76      unsigned int Hun,Ten;
77
78      Hun = distanceInt/100;
79      distanceChar[0] = (unsigned char) Hun;
80      distanceChar[0] += 0x30;
81       distanceInt = distanceInt % 100;
82
83       Ten = distanceInt/10;
84      distanceChar[1] = (unsigned char) + Ten;
85      distanceChar[1] += 0x30;
86      distanceInt = distanceInt % 10;
87
88      distanceChar[2] = (unsigned char)distanceInt;
89      distanceChar[2] += 0x30;
90   }
91   void calculate_Display(void){
92
93      sendCmd(ClrDisp);
94       sendData('D');
95       sendData(' ');
96       sendData('=');
97       sendData(' ');
98
99       distanceInt /= 58          // 58 counts per cm
100      intToascii()               // convert to string
101
102
103     sendData(distanceChar[0]);
104     sendData(distanceChar[1]);
105     sendData(distanceChar[2]);
106   }
107
108
109   void main(void){
110   // f=4MHz(internal clock), timer1 cycle = 1MHz, 1uS
111      unsigned int i;
112
113      TMR1CS  = 0;                // set as timer
114      T1CKPS0 = 0;                // prescale 1:1
115      T1CKPS1 = 0;
```

Listing 6.3   (*continued*)

```
116
117     CMCON = 0x07;          // RA0,1,2 as digital I/Os
118     TRISA = 0xB8;          // set inputs/outputs
119     TRISB = 0x00;
120     RW = 0;                // Write mode
121     RS = 0;                // Command mode
122     E = 0;                 // Start talking to LCD
123
124     initializeLCD();
125     sendData('P');
126     sendData('r');
127     sendData('e');
128     sendData('s');
129     sendData('s');
130     sendData(' ');
131     sendData('t');
132     sendData('o');
133     sendData(' ');
134     sendData('M');
135     sendData('e');
136     sendData('a');
137     sendData('s');
138     sendData('u');
139     sendData('r');
140     sendData('e');
141
142     i = 0;
143     distanceInt = 0;
144     TMR1H = 0;             // timer reload values
145     TMR1L = 0;
146
147   while(1){
148     while(Measure)        // wait for measurement command
149       ;
150     Tx = 1;
151     delay_uS(100);
152     Tx = 0;
153       TMR1ON=1;           // Rx received, timer ON
154     while(Rx==0)
155       ;
156     TMR1H = 0;            // timer reload values
157     TMR1L = 0;
158     while(Rx == 1)        // wait for Rx pulse to finish
159       ;
160     TMR1ON=0;            // Rx finished, timer OFF
161     distanceInt = ((TMR1H*256) + TMR1L);
162     calculate_Display();
163     delay_mS(10);         // time required after the Rx
164     }
165 }
```

Listing 6.3 (*continued*)

Most of the program in Listing 6.3 has already been discussed in Chapter 3 but some parts such as 'intToascii()', 'calculate_Display()' and the 'main()' functions are new and are therefore explained in more detail in this section. The 'main()' function starts at line 109 and performs the following tasks:

- timer configuration
- LCD initialization
- I/O lines configuration
- sending the message 'Press to Measure' and display of this message on the LCD
- generation of the pulse for the ultrasonic module's transmitter
- measurement of the returned pulse and calculation of corresponding distance.

After the calculation of the measured distance at line 161, the function 'main()' calls the 'calculate_Display()' function.

The 'calculate_Display()' function first clears the LCD display, then sends a message of 'D ='. We display the distance in centimetres so the timer count is divided by 58 at line 99. This conversion factor has been obtained by making experimental measurements of the ultrasonic module used in the circuit (Ultrasonic range finder SRF04) and taking into consideration the various system parameters such as timer clock. The LCD can only display characters, so at line 100 the 'intToascii()' function is called, which converts an integer value to a string of characters and places them into the string buffer 'distanceChar[]'. Finally, the values of the buffer are sent one at a time to the LCD to be displayed.

The 'intToascii()' function divides the integer number into three characters of Hundreds (Hun), Tens (Ten) and Ones. At line 78 the number of hundreds in the integer is calculated, which at line 79 is casted (forced) to be character type. To make sure this ASCII number is displayed as a number, a value of 30 hex is added to it and then it is placed into the character buffer.

Line 81 calculates the modulus of a number, which means the remainder. For example, if the distanceInt value is 267 here, then after the execution of line 81, distance would be 67.

The Tens are calculated in the same manner at lines 83–86. The Ones must just be converted to ASCII, as this has already been calculated at line 86.

To further develop this project, the reader could modify the program in order to perform the measurement in inches as well as centimetres, measure two distances and calculate the area, or measure three distances and calculate the volume.

6.5 Function Generator

This project is concerned with the design of a simple signal generator that can generate and output one of three basic waveforms – sine, square and triangular. To simplify the hardware design procedure we use the circuit developed in the previous chapter with slightly modified pin functions. The desired waveform can be selected by pressing one of three buttons attached to three pins of PORTA: pin 0 for square wave, pin 1 for triangular wave and pin 2 for sine wave. Other signal types can be easily added by modifying this basic program. The digital signal goes to PORTB, where a network of resistors converts it into analog form as explained in Chapter 5. The final analog signal therefore has an 8-bit resolution. The pins of PORTC are unused for this project, so resolution of the analog signal can be increased to 10 or 12 bits by using two or four bits from PORTC. This would require some minor program modifications as well as some hardware modifications and the addition of more resistors to the existing 8-bit R–2R ladder network. The Timer 2 interrupt is used to generate the sampling frequency for the output signal. It is left to readers to determine this sampling frequency from the initialization of registers related to the operation of Timer 2, as described in Chapter 5. The full program is shown in Listing 6.4.

```
1   // this program can generate three different
2   // types of signals on port B: square, sine and triangular wave
3   // sampling rate: xxx
4   // frequency range for signals: xxx
5   // type of signal and its frequency are selected
6   // through the pins 0-3 of port A
7
8   #include <p18F252.h>
9
10  #define pi 3.14159265359
11  #define Ns 80          // number of samples per period
12  // for lowest frequency available
13  #define fs 4000        // sampling frequency
14  #define f 50           // starting frequency
15
16  // Global Variables
17  unsigned char i = 0;
18  unsigned char SQ[Ns];
19  unsigned char TR[Ns];
20  unsigned char SI[Ns];
```

Listing 6.4   Signal generator using 8-bit DAC

```
21
22   // counter and signal type selector
23   unsigned char count, signal_type;
24
25   void timer_isr (void);
26
27   #pragma code low_vector=0x10
28   void low_interrupt (void)
29   {
30      _asm GOTO timer_isr _endasm
31   }
32
33   #pragma code
34
35   #pragma interruptlow timer_isr
36
37   void timer_isr(void)
38   {
39      unsigned char temp;
40
41   if (signal_type == 1) temp = SQ[count];
42   if (signal_type == 2) temp = TR[count];
43   if (signal_type == 3) temp = SI[count];
44
45      PORTB = temp; // output sample
46
47        count = count+1;           // increment counter
48   if (count >= Ns)
49        count = 0;                 // and reset when needed
50
51   PIR1bits.TMR2IF = 0;
52   }
53
54   void Timer2_Init(void)
55   {
56      T2CON = 0x4D;
57      PR2 = 0x18;
58   PIE1bits.TMR2IE = 1;
59   INTCONbits.PEIE = 1;
60   }
61
62   void Lookup_Init(void)
63   {
64   unsigned char i;
65   float temp;
66
67   // initialize all three lookup tables
68   for (i=0; i<Ns; i++)
69      {
70        if (i<Ns/2)
71        {
72           SQ[i]=255;
73           TR[i]=(i*510)/Ns;
```

Listing 6.4   (*continued*)

```
74            }
75            if(i>=Ns/2)
76            {
77                SQ[i]=0;
78                TR[i]=((Ns-i)*510)/Ns;
79            }
80            temp = (1+sin((2*pi*f*i)/fs))*127;
81            SI[i] = (int)temp;
82        }
83    }
84
85    void wait(void)
86    {
87    unsigned int x,y;
88
89        // roughly 1s delay for pushbutton press checks
90        for (y=0; y<100; y++)
91            for (x=0; x<2650; x++) {}
92    }
93
94    void main (void)
95    {
96    float alarm_level = 1.;
97    unsigned char temp;
98
99        ADCON1 = 0x07;          // portA is digital
100       TRISA = 0xFF;           // portA is input
101       TRISB = 0x00;           // portB is output
102
103       count = 0;
104
105   signal_type = 1;            // start with the square wave
106                   // 1 = SQ, 2 = TR, 3 = SI
107
108   Lookup_Init();             // initialize lookup tables
109    Timer2_Init();                    // set the timer
110
111   INTCONbits.GIE = 1;    // enable interrupts
112
113   while(1)      // wait for interrupts
114   {
115      wait();
116
117   temp = PORTA;        // read port A to select signal
118               // and change signal frequency if needed
119
120      if(temp != 0x00)
121   {
122        // signal type selection – pins RA0, RA1 and RA2

123        if((temp & 0x01) == 0x01) // select SQ wave signal
124            signal_type = 1;
125        if((temp & 0x02) == 0x02) // select TR wave signal
126            signal_type = 2;
```

Listing 6.4 (*continued*)

```
127    if((temp & 0x04) == 0x04) // select SI wave signal
128      signal_type = 3;
129    }
130    // if temp == 0, no selection made, keep going
131  }
132  }
```

Listing 6.4 (*continued*)

The signal-generation approach adopted in this program involves precalculating the samples for each of three waves and saving them into microcontroller RAM memory. The program then circulates through the relevant parts of the memory and outputs samples to generate the selected signal waveform. This saves some computation time, especially in the case of the sine wave where the sin() function needs to be called to calculate each sample of the wave. The downside of this approach is that a large part of microcontroller RAM memory needs to be reserved for the storage of those waveforms, which can limit the time resolution of generated signals. Square and triangular waves require much less computational effort so this approach is not necessary for those signals, but they have also been pre-computed to make the program easier to follow and understand. This has been done in the function Lookup_Init(void), defined between lines 62 and 83. Three lookup tables – arrays of 80 elements, declared at the beginning of the program (lines 18, 19 and 20) – have been filled with appropriate values in order to generate a single period of each signal type. Square wave array, SQ[] holds value 255 –the maximum 8-bit value in the first half of the array (40 elements), and the minimum value, 0, in the other 40 elements. The first 40 elements of the triangular wave array, TR[] linearly increase in value from 0 to 255. The next 40 elements linearly decrease from that value back to 0. Finally, elements of the sine wave array SI[] are generated using the sine wave equation, implemented in line 80 of the program. The sin() function used in this equation produces a result in double format. This result needs to be converted to integers before being stored into the SI[] array, which is declared as a character-type array in line 20 of this program. This is done using type casting in line 81 of the program. Notice also that the value of π needs to be defined for the call of function sin() in line 80. This constant, pi, has been defined in line 10.

Pins 0, 1 and 2 of PORTA are used to select the signal type. The value of those pins is checked in the main part of the program between lines 120 and 129. To check the value of each of those three bits from PORTA, the state of this port is read and stored into variable 'temp', which is later logically 'and-ed' with values 0x01, 0x02 and 0x04. This technique, called masking, is used to reset the value of all other bits of variable 'temp' and reveal the value of the relevant bit for our application. If any of the three tested bits is found to be 1, the value of the signal_type variable is modified to reflect this.

The signal sample is the output to PORTB in the Timer2 interrupt service routine, called whenever Timer2 overflows, creating a constant sampling rate for our signal generator. The signal sample is output from one of three arrays according to the value of the signal_type variable. In this way the selection of signal type can be achieved by setting one of three least significant bits of PORTA. It is recommended to attach a push button to each of those pins and label each button appropriately to enable easier operation of the final system. It is also recommended that readers try to modify the program further in order to avoid the switch-bouncing problem encountered when push-button switches, toggle switches or electro-mechanical relays are used in microcontroller-based systems. These devices usually have metal contacts that can close or open the circuit and carry the current in switches and relays. The contacts are made of metal so they have mass and at least one of the contacts is on a movable strip of metal, so it has springiness. These contacts usually have to be designed to open and close quickly so there is little damping in their movement and they will be 'bouncy' as they close and open. When an open pair of contacts is closed, the contacts will come together and bounce off each other several times before finally coming to rest in a closed position. The effect is called 'contact bounce' or, in a switch, 'switch bounce'. In most cases contacts can bounce on opening as well as on closing. A simplified time history of switch opening and closing is shown in Figure 6.5.

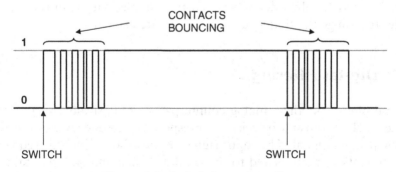

Figure 6.5 Switch bouncing effect

Contact bouncing can cause problems in an application like ours as an open switch can be read as closed due to a bouncing effect and vice versa. The simplest way to try to prevent this problem is to ensure that only one reading of the switch or button is attempted in a given time period (for example, 0.5 s). We have implemented this approach in our program by inserting a delay of approximately 1 s before reading PORTA in the main loop of the program. A simple wait() function is defined between lines 85 and 92 and called in line 115 of the main program. The reader is invited to experiment with this delay function or to extend and improve this software debouncing approach further. A function that reads the switch pin a number of times and only returns a stable value after a number of same reading results can be considered. Alternatively, an even simpler approach can be to read the state of the switch after each complete cycle of the output signal.

This debounce routine must be tuned to a particular system and application and you should also be aware that switches and relays can lose some of their springiness as they age. That can cause the time it takes for contacts to stop bouncing to increase with time. So the debounce function that worked fine when the switch was new might not work a year or two later.

As a good exercise and an additional feature of the function generator created in this section, we recommend adding a frequency adjustment feature to an existing system. An additional two pins of PORTA can be used to increase or decrease the signal frequency. The state of these pins would then have to be detected by the program and an additional variable added to the program to increment an array counter by a number greater than 1 (as currently implemented in line 47 of the program). In this way, the program would go through the array skipping a number of elements from the array, thus increasing the frequency of the output signal. The downside of this approach would be the reduced resolution of the output signal. Alternatively a delay between the samples can be introduced in order to change the frequency of the generated signal.

6.6 Digital Filtering

Digital filters, like their analog counterparts, are basically signal conditioners. They transmit or reject a (prespecified) frequency range of the original input signal. The input signal, minus those blocked frequency components, is then passed to the output. While analog filters take an analog input and produce an analog output, digital filters take a digital input and produce a digital output. In a typical digital filtering application, a digital filtering program running on a microprocessor system reads the input samples from an AD converter, performs the mathematical manipulations dictated by filtering theory for the required filter type and specification and outputs the result via a DAC back into the analog world.

There are many filter types but the most common ones, classified according to the frequency range that they reject or transmit, are lowpass, highpass, bandpass and bandstop filters. A lowpass filter, for example, allows only low-frequency signals (below some specified cutoff) through to its output while a highpass filter does just the opposite, by rejecting only frequency components below some threshold. Bandpass and bandstop filters can block or pass the range of frequencies specified with two cutoff points.

An analog filter operating directly on the analog inputs consists of analog components – resistors, capacitors, inductors and analog amplifiers. A digital filter combines an entirely different set of components to process the digital input signal – digital delays, multipliers and adders. Those components can be combined to result in a digital filter with very complex frequency responses.

We will skip, here, all of the difficult filter theory on how to combine those elements in order to achieve the desired frequency response from the filter and will concentrate instead on the actual implementation of a filter using a PIC microcontroller. It needs to be stressed here that a PIC microcontroller is not the most suitable device to implement a digital filtering system due to the very high computational demands of most filtering algorithms. A dedicated group of microprocessors, digital signal processors (DSPs), has been developed during the last couple of decades and used for these tasks but for simple digital filters with a small number of coefficients a general-purpose PIC MCU can still be used successfully. The two main types of digital filters with respect to mathematical equations used to implement them are finite impulse response (FIR) filters and infinite impulse response (IIR) filters. We will implement both FIR and IIR digital filters using the PIC18F252-based system discussed in Chapter 5. This circuit is suitable for this application as we have already explained and tested the basic 'talkthrough' program that takes the analog signal from the microcontroller input, performs conversion into its digital form equivalent and passes each sample through the microcontroller and outputs unmodified signal to the eight bits of PORTB. The resistor ladder network attached to PORTB, acting as a digital to analog converter, converts the output signal back into analog domain. In order to implement FIR- based digital filtering on this system, all that needs to be done is to add an FIR filtering function to it. This function needs to implement the FIR equation given below.

$$y(n) = h_0 \cdot x(n) + h_1 \cdot x(n-1) + h_2 \cdot x(n-2) + \dots$$
$$+ h_{N-1} \cdot x(n-(N-1))$$

which can be represented as:

$$y(n) = \sum_{k=0}^{k=N-1} h_k \cdot x(n-k)$$

In the above equations:

y(n) is the output of the digital filter at sampling instant n
x(n) is the input to the digital filter at sampling instant n
$x(n-1)$ is the input to the digital filter at sampling instant $n-1$, or more generally
$x(n-\text{k})$ is the input to the digital filter at sampling instant $n-k$
$h_0, h_1 \dots h_{N-1}$ are constants specific for the particular digital filter – the filter coefficients.

The FIR filter therefore simply produces a weighted average of its N most recent input samples. All of the magic is in the coefficients, which dictate the actual output for a given sequence of input samples.

The length of this filter is N as it has N filter coefficients ($h_0, h_1, \dots, h_{N-1}$). The process of selecting the filter's length and coefficients is called filter design. The goal of this process is to set those parameters such that certain desired stopband and passband parameters will result from running the filter. Most engineers use dedicated mathematical programs such as Matlab to perform filter design. The longer the filter (more coefficients) the more finely its response can be tuned, but the computational load for implementing such a filter increases rapidly and may limit the length of the filter if a relatively high sampling frequency needs to be realized.

Figure 6.6 shows the basic block diagram for an FIR filter of length N. Note that this diagram essentially represents a graphical illustration of the FIR equation given earlier. Boxes labelled 'delay' in this diagram denote a delay of the input signal by one sampling period.

The IIR type filters are slightly more complex as they also use feedback (previous filter outputs) to produce a new filter output. This 'historical' information helps IIR filters to achieve finely tuned frequency responses with a smaller number of coefficients and therefore less computational effort compared to FIR filters. The basic IIR filter equation is given

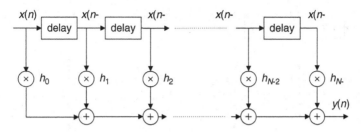

Figure 6.6 Structure of FIR digital filter

below:

$$y(n) = b_0 \cdot x(n) + b_1 \cdot x(n-1) + b_2 \cdot x(n-2) + \ldots$$
$$+ b_{N-1} \cdot x(n-(N-1)) - a_1 \cdot y(n-1)$$
$$+ a_2 \cdot y(n-2) + \ldots + a_M \cdot x(n-M)$$

abbreviated as:

$$y(n) = \sum_{k=0}^{k=N-1} b_k \cdot x(n-k) - \sum_{k=1}^{k=M} a_k \cdot y(n-k)$$

Here:

$y(n)$ is the output of the digital filter at sampling instant n
$y(n-1), y(n-2), \ldots, y(n-M)$ are M old filter outputs
$x(n)$ is the input to the digital filter at sampling instant n
$x(n-1), x(n-2), \ldots, x(n-N+1)$ are $N-1$ old filter inputs
b_0, b_1 and b_{N-1} and a_1, a_2 and a_M are two sets of IIR filter coefficients.

Coefficients a_1, a_2 and a_M multiply old filter outputs that are fed back into the digital filter, so those coefficients are also sometimes called 'feedback filter coefficients'. Coefficients b_0, b_1 and b_{N-1} are known as 'feedforward filter coefficients'.

A block diagram implementing the IIR digital filter is shown in Figure 6.7. Note that this diagram illustrates a specific case where $M = N$. This does not have to be the case for most IIR filters.

Programs implementing FIR and IIR digital filters are given in Listing 6.5 and Listing 6.6. Coefficients for both filters are designed to realize low-pass filter with cutoff frequency of 100 Hz. Two sets of filter coefficients are suggested for both filters, realizing filters of different types and order. By removing comments from the lines specifying the second set of filter coefficients in each program and

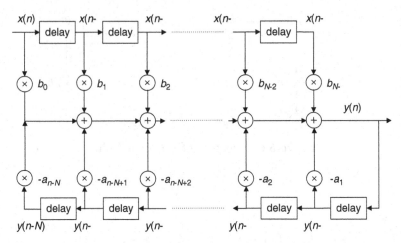

Figure 6.7 Structure of the IIR digital filter

commenting out (i.e. using '/* */' C comment signs around) the current set of filter coefficients, readers can test both sets of filters and compare performance. Techniques used to design those filters and obtain filter coefficients are beyond the scope of this book. Both programs are using the same 'talkthrough' template program discussed in Chapter 5. The only significant addition is the FIR filtering function in Listing 6.5 and IIR filtering function in Listing 6.6. It is left to readers to analyse and test those functions and to understand how the equations given in this chapter have been implemented into C functions 'fir()' (Listing 6.5) and 'iir()' (Listing 6.6). We recommend the using the MPLAB- Simulator and suitable stimulus files to step slowly through each of these two functions and programs in order to understand operations of both filtering operations more easily. Appropriate comments have been made for each program, which should also help readers to understand its operation.

Finally, reader can try to 'measure' the frequency response of both implemented filters. By dividing the amplitude of output and input signal for a sine wave signal of certain frequency, a value of filter response at that frequency can be obtained. 'Sweeping' through the frequency range of interest and performing similar measurements using MPLAB, a frequency response can be obtained for each filter. Once implemented on the circuit described in Chapter 5, the frequency response of the filter operating in real time can be measured using a signal generator to generate sine waves of various frequencies and an oscilloscope to measure the amplitude of the filter output.

```
 1  // this program can perform LP FIR type filtering
 2  // on the input signal from AN0 pin
 3  // sampling frequency set to 1 kHz
 4  // low-pass filter cutoff frequency is 100 Hz
 5
 6  #include <p18F252.h>
 7
 8  #define Mfir 4          // FIR filter order
 9
10  // Global Variables
11  unsigned char oldest = 0;
12  unsigned char output = 0;
13  float sample;
14
15  // ********** FIR filter arrays **********
16  float u[Mfir+1];        // FIR delay buffer
17  // FIR coefficients - 10 coefficients
18  /* float h[Mfir+1] = {0.0218584809574104,
19          0.0735766021143549,
20          0.127323954473516,
21          0.171678738266828,
22          0.196726328616693,
23          0.196726328616693,
24          0.171678738266828,
25          0.127323954473516,
26          0.0735766021143549,
27          0.0218584809574104};
28  */
29  // FIR coefficients - 5 coefficients
30  float h[Mfir+1] = {0.1513653,
31          0.1870979,
32          0.2,
33          0.1870979,
34          0.1513653};
35
36  void timer_isr (void);
37  float getSample(void);
38  float fir(unsigned char M, float *h, float *w, float x);
39  void buff_reset(void);
40
41  #pragma code int_vector=0x08
42  void low_interrupt (void)
43  {
44  _asm GOTO timer_isr _endasm
45  }
46
47  #pragma code
48
49  #pragma interrupt timer_isr
50
```

Listing 6.5   FIR filtering program for PIC18F252

```
51  void timer_isr(void)
52  {
53  unsigned char i, output;
54  float y;
55
56  sample = getSample(); // get new sample from the ADC
57
58  y = fir(Mfir, h, u, sample); // do the FIR filtering
59
60  output = (int)(y*51); // scale and convert to int/char
61  // scaling factor for 8 bit output = 255/5 = 51
62
63  PORTB = output;          // output result
64
65  PIR1bits.TMR2IF = 0; // clear the interrupt flag
66  }
67
68  void Timer2_Init(void)
69  {
70  T2CON = 0x4F;
71  PR2 = 0x18;
72  PIE1bits.TMR2IE = 1;
73  INTCONbits.PEIE = 1;
74  }
75
76  float getSample(void)
77  {
78      ADCON0bits.GO = 1;              //start a conversion
79      while (ADCON0bits.GO == 1); // wait for completion
80      return((ADRESH * 256 + ADRESL) * 0.00489);
81  }
82
83  // ********** FIR filtering function y = fir(M, h, u, x)
84  // M - filter length
85  // h - coefficients
86  // u - delay buffer
87  // x - new input sample
88  float fir(unsigned char M, float *h, float *u, float x)
89  {
90      unsigned char i;      // internal counter
91      float y;       // filter output to be calculated
92
93      u[0] = x; // read new input sample
94
95      for (y=0, i=0; i<=M; i++)
96          y += h[i] * u[i]; // calculate output
97
98      for (i=M; i>=1; i--) // update delay buffer
99          u[i] = u[i-1]; // done in reverse order
100
101          return y;
```

Listing 6.5   (*continued*)

```
102  }
103
104  // reset delay buffer for FIR filter
105  void buff_reset (void)
106  {
107      int i;
108
109    for (i = 0; i < Mfir; i++) u[i] = 0.0; // FIR delay buffer
110  }
111
112  void main (void)
113  {
114      float alarm_level = 1.;
115
116      ADCON0 = 0x81;      // select input AN0, enable ADC
117      ADCON1 = 0x80;      // RA0 is analog
118      TRISA = 0x01;       // RA0 is input
119      TRISB = 0x00;       // portB is output
120      Timer2_Init();      // set the timer
121      INTCONbits.GIE = 1; // enable interrupts
122
123      buff_reset();
124      while(1); // wait for interrupts
125  }
```

Listing 6.5 (*continued*)

```
1   // this program can perform LP IIR type filtering
2   // on the input signal from AN0 pin
3   // sampling frequency set to 1 kHz
4   // low-pass filter cutoff frequency is 100 Hz
5
6   #include <p18F252.h>
7
8   #define Miir 2      // IIR filter - numerator order
9   #define Liir 2      // IIR filter - denominator order
10
11  // Global Variables
12  unsigned char oldest = 0;
13  unsigned char output = 0;
14  float sample;
15
16  // ********** IIR arrays **********
17  float w[Miir+1];        // IIR numerator delay buffer
18  float v[Liir+1];        // IIR denominator delay buffer
19  // IIR numerator coefficients - Butterworth type,
    2nd order
20  float b[Miir+1] = {0.06745527,
21         0.13491055,
22         0.06745527};
```

Listing 6.6 IIR filtering program for PIC18F252

```
23  // IIR denominator coefficients
24  float a[Liir+1] = {1.0000000,
25           -1.1429805,
26           0.4128016};
27
28  // IIR numerator coefficients - Bessel type filter, 3rd order
29  /* float b[Miir+1] = {0.018098933,
30           0.054296799,
31           0.054296799,
32           0.018098933};
33  // IIR denominator coefficients
34  float a[Liir+1] = {1.0000000,
35           -1.7600419,
36           1.1828933,
37           -0.27805992};
38  */
39
40  void timer_isr (void);
41  float getSample(void);
42  float iir(unsigned char M, float *a, unsigned char L,
    float *b, float *w, float *v, float x);
43  void buff_reset(void);
44
45  #pragma code int_vector=0x08
46  void low_interrupt (void)
47  {
48  _asm GOTO timer_isr _endasm
49  }
50
51  #pragma code
52
53  #pragma interrupt timer_isr
54
55  void timer_isr (void)
56  {
57  unsigned char i, output;
58  float y;
59
60  sample = getSample(); // get new sample from the ADC
61
62  y = iir(Miir, a, Liir, b, w, v, sample); // IIR filtering
63
64  output = (int)(y*51); // scale and convert to int/char
65  // scaling factor for 8 bit output = 255/5 = 51
66
67  PORTB = output;         // output result
68
69  PIR1bits.TMR2IF = 0; // clear the interrupt flag
70  }
```

Listing 6.6   (*continued*)

```
71  72  void Timer2_Init(void)
73  {
74  T2CON = 0x4F;
75  PR2 = 0x18;
76  PIE1bits.TMR2IE = 1;
77  INTCONbits.PEIE = 1;
78  }
79
80  float getSample(void)
81  {
82  ADCON0bits.GO = 1;          //start a conversion
83  while (ADCON0bits.GO == 1);       // wait for completion
84  return((ADRESH * 256 + ADRESL) * 0.00489);
85  }
86
87
88  //********** IIR filtering function y = iir (M,a,L,b,w,v,x)
89  // M, L - denominator and numerator orders
90  // a, b - denominator and numerator coefficients
91  // v, w - delay buffers
92  // x - new input sample
93  float iir(unsigned char M, float *a, unsigned char L,
    float *b, float *w, float *v, float x)
94  {
95      unsigned char i;
96    float y = 0;
97
98      v[0] = x;               // read new input sample
99      w[0] = 0;            // filter output to be calculated
100
101     for (i=0; i<=L; i++) // numerator calculation
102           w[0] += b[i] * v[i];
103
104     for (i=1; i<=M; i++) // denominator calculation
105           w[0] -= a[i] * w[i];
106
107     for (i=L; i>=1; i--) // delay buffers updating
108           v[i] = v[i-1];      // in reverse order
109
110     for (i=M; i>=1; i--)
111           w[i] = w[i-1];
112
113     return w[0];            // new output sample
114  }
115
116  // reset delay buffers for FIR and IIR filters
117  void buff_reset(void)
118  {
119  int i;
```

Listing 6.6   (*continued*)

```
120
121  for (i = 0; i < Miir; i++)  w[i] = 0.0; // IIR numerator
122  for (i = 0; i < Liir; i++)    v[i] = 0.0; // IIR denominator
123  }
124
125  void main (void)
126  {
127  float alarm_level = 1.;
128
129  ADCON0 = 0x81;           // select input AN0, enable ADC
130  ADCON1 = 0x80;           // RA0 is analog
131  TRISA = 0x01;            // RA0 is input
132  TRISB = 0x00;            // portB is output
133  Timer2_Init();           // set the timer
134  INTCONbits.GIE = 1;      // enable interrupts
135
136  buff_reset();
137  while(1);                // wait for interrupts
138  }
```

Listing 6.6 (*continued*)

Appendix

More information related to projects discussed in this book can be found in other books or on various Internet sites. We list here a few that can be useful for readers trying to access and download free versions of the software used in this book or who are looking for more information related to some of the projects described.

Web Sites

Microchip home page: http://www.microchip.com/, accessed 29 December 2008 (for more information on PIC microcontrollers, data sheets, the latest version of MPLAB software and C18 compiler).

HI-TECH page: http://www.htsoft.com/, accessed 29 December 2008 (for free HI-TEC compiler for PICs).

CadSoft page: http://www.cadsoftusa.com/, accessed 29 December 2008 (to download Eagle software for drawing schematics and PCB design).

Books

Further information on stepper and other DC motors:

Polka, D. (2002) *Motors and Drives: A Practical Technology Guide*, The Instrumentation, Systems, and Automation Society, Triangle Park NC.

Further information on digital filters design and coefficients calculation:

Ingle, V.K. and Proakis, J.G. (2006) *Digital Signal Processing Using Matlab*, 2nd edn, Thomson Learning, Boston MA.

Index

Note: Page references in *italics* refer to Figures and Tables